Thermophiles
Biodiversity, Ecology, and Evolution

Thermophiles
Biodiversity, Ecology, and Evolution

Edited by

Anna-Louise Reysenbach
Portland State University
Portland, Oregon

Mary Voytek
United States Geological Survey
Reston, Virginia

and

Rocco Mancinelli
NASA/Ames Research Center
Moffett Field, California

Springer Science+Business Media, LLC

Library of Congress Cataloging-in-Publication Data

Reysenbach, Anna-Louise; Voytek, Mary; Mancinelli, Rocco
Thermophiles: Biodiversity, Ecology, and Evolution/Anna-Louise Reysenbach, Mary Voytek, Rocco Mancinelli
p. cm.
Includes bibliographical references and index.
ISBN 978-1-4613-5436-9 ISBN 978-1-4615-1197-7 (eBook)
DOI 10.1007/978-1-4615-1197-7
1. 2.

ISBN 978-1-4613-5436-9

Originally published by Kluwer Academic/Plenum Publishers, New York in 2001
Softcover reprint of the hardcover 1st edition 2001
http://www.wkap.nl/

10 9 8 7 6 5 4 3 2 1

A C.I.P. record for this book is available from the Library of Congress

In memory of Rick Hutchinson and chocolate

There are only meters in this book,
and it took more than 365 days
to put the book together.

Contributors

Kai S. Anderson, *Department of Geological and Environmental Sciences, Stanford University, Stanford, California 94305-2115*

Mary M. Bateson, *Department of Microbiology, Montana State University, Bozeman, Montana 59717*

Deborah A. Body, *School of Biological Sciences, University of Wales, Bangor, LL57 2UW, Wales*

Deena Braunstein, *Department of Geological and Environmental Sciences, Stanford University, Stanford, California 94305-2115*

Toni A. M. Bridge, *School of Biological Sciences, University of Wales, Bangor, LL57 2UW, Wales*

Thomas D. Brock, *E. B. Fred Professor of Natural Sciences Emeritus, University of Wisconsin-Madison, Madison, Wisconsin 53705*

C. K. Browning, *Ransom Hill Bioscience, Inc., Ramona, California 92065*

Debby F. Bruhn, *Lockheed Martin Idaho Technologies Co., Idaho National Engineering and Environmental Laboratory, Idaho Falls, Idaho 83415-2203*

D. K. Bulmer, *Biotechnologies Department, Idaho National Engineering and Environmental Laboratory, Idaho Falls, Idaho 83415-2203*

Siegfried Burggraf, *Lehrstuhl für Mikrobiologie and Archaeenzentrum, Universität Regensburg, 93053 Regensburg, Germany*

Charles C. Chester, *World Foundation for Environment and Development, Washington, D.C. 20036*

Joan Combie, *Montana Biotech, Belgrade, Montana 59714*

M. J. Ferris, *Department of Microbiology, Montana State University, Bozeman, Montana 59717*

Joseph R. Graber, *Department of Microbiology and Biochemistry, Cook College, Rutgers University, New Brunswick, New Jersey 08903*

Robert Huber, *Lehrstuhl für Mikrobiologie and Archaeenzentrum, Universität Regensburg, 93053 Regensburg, Germany*

Christian Jeanthon, *CNRS, UPR 9042 and UPMC, Station Biologique, Roscoff and Université de Bretagne Occidentale, Brest, France*

D. Barrie Johnson, *School of Biological Sciences, University of Wales, Bangor, LL57 2UW, Wales*

Julie Kirshtein, *Department of Microbiology and Biochemistry, Cook College, Rutgers University, New Brunswick, New Jersey 08903*

Jan W. de Leeuw, *Division of Marine Biochemistry, Netherlands Institute for Sea Research (NIOZ), 1790 AB Den Burg, Texel, Netherlands*

Robert F. Lindstrom, *Yellowstone Center for Resources, Yellowstone National Park, Wyoming 82190*

Donald R. Lowe, *Department of Geological and Environmental Sciences, Stanford University, Stanford, California 94305-2115*

M. T. MacDonell, *Ransom Hill Bioscience, Inc., Ramona, California 92065*

Michael T. Madigan, *Department of Microbiology, Southern Illinois University, Carbondale, Illinois 62901*

Thomas Mayer, *Lehrstuhl für Mikrobiologie and Archaeenzentrum, Universität Regensburg, 93053 Regensburg, Germany*

S. C. Nold, *Montana State University, Department of Microbiology, Bozeman, Montana 59717*

William D. O'Dell, *Department of Biochemistry and Molecular Biology, University of Nebraska Medical Center and Department of Biology, University of Nebraska at Omaha, Omaha, Nebraska 68198-4525*

Daniel Prieur, *CNRS, UPR 9042 and UPMC, Station Biologique, Roscoff and Université de Bretagne Occidentale, Brest, France*

Reinhard Rachel, *Lehrstuhl für Mikrobiologie and Archaeenzentrum, Universität Regensburg, 93053 Regensburg, Germany*

Robert F. Ramaley, *Department of Biochemistry and Molecular Biology, University of Nebraska Medical Center and Department of Biology, University of Nebraska at Omaha, Omaha, Nebraska 68198-4525*

Anna-Louise Reysenbach, *Department of Environmental Biology, Portland State University, Portland, Oregon 97201*

Francisco F. Roberto, *Biotechnologies Department, Idaho National Engineering and Environmental Laboratory, Idaho Falls, Idaho 83415-2203*

Petra Rossnagel, *Lehrstuhl für Mikrobiologie and Archaeenzentrum, Universität Regensburg, 93053 Regensburg, Germany*

Lynn J. Rothschild, *Ecosystem Science and Technology Branch, NASA/Ames Research Center, Moffet Field, California 94035-1000*

Kenneth Runnion, *Montana Biotech, Belgrade, Montana 59714*

C. M. Santegoeds, *Montana State University, Department of Microbiology, Bozeman, Montana 59717*

Pamela L. Scanlan, *Department of Biochemistry and Molecular Biology, University of Nebraska Medical Center and Department of Biology, University of Nebraska at Omaha, Omaha, Nebraska 68198-4525*

Mark Speck, *Department of Microbiology and Biochemistry, Cook College, Rutgers University, New Brunswick, New Jersey 08903*

D. L. Stoner, *Biotechnologies Department, Idaho National Engineering and Environmental Laboratory, Lockheed Martin Idaho Technologies Co., Idaho Falls, Idaho 83415-2203*

John D. Varley, *Yellowstone Center for Resources, Yellowstone National Park, Wyoming 82190*

Mary Voytek, *United States Geological Survey, MS430, Reston, Virginia 20192*

David M. Ward, *Department of Microbiology, Montana State University, Bozeman, Montana 59717*

T. E. Ward, *Biotechnologies Department, Idaho National Engineering and Environmental Laboratory, Idaho Falls, Idaho 83415-2203*

Preface

These are indeed exciting times to be a microbiologist. With one of the buzzwords of the past decade—"Biodiversity," and microbes are reveling in the attention as they represent by far most of the biodiversity on Earth. Microbes can thrive in almost any environment where there is an exploitable energy source, and, as a result, the possible existence of microbial life elsewhere in the solar system has stimulated the imaginations of many. Extremophiles have taken center stage in these investigations, and thermophiles have taken on the lead roles. Consequently, in the past decade there has been a surge of interest and research in the Ecology, Biology, and Biotechnology of microorganisms from thermal environments. Many of the foundations of thermophile research were laid in Yellowstone National Park, primarily by the research of Professor Thomas Brock's laboratory in the late 1960s and early 1970s. The upper temperature for life was debated, the first thermophilic archeum discovered (although it was only later shown to be an archeum by ribosomal cataloging), and the extremes of light, temperature, pH on the physiology of microorganisms were explored. Interest in thermophiles increased steadily in the 1970s, and with the discovery of deep-sea hydrothermal vents in 1977, thermophilic research began its exponential explosion. The development of *Taq* polymerase in the polymerase chain reaction (PCR) focused interest on the biotechnological potential of thermophilic microorganisms and on the thermal features in Yellowstone National Park. Additionally, the use of *Taq* polymerase in molecular phylogenetic approaches to assess microbial diversity has identified a plethora of novel types of microbes representing a diverse array of potential metabolic types.

This book aims to provide a source of the recent advances in the biology, biotechnology, and management of thermophilic microorganisms. The contributed chapters include research results, technical information, and reviews that highlight the state of our current knowledge of thermophiles and their habitats. Most of the contributors have drawn specific examples largely from the Yellowstone National Park thermal springs. The volume presents a historical background and gives an overview of research on thermophilic microorganisms (Brock; Prieur et al.). Section II covers what is known about the microbial diversity associated with hydrothermal environments and addresses the unusual physiologies of some of these organisms. Estimates of the diversity of natural microbial communities have been hindered by our inability to enrich and culture the organisms present (Ferris et al.). Yet such enrichment procedures established the foundations of what we know of their diversity and physiology (Ramaley et al.; Johnson et al.). Recent advances in molecular phylogenetic techniques have eliminated the need to culture organisms to assess diversity. Studies based on these techniques have yielded a plethora of diverse, and, sometimes, novel lineages, such as the Korarchaeota (Graber et al.; Ferris et al.; Stoner et al.). Section III contains papers pertaining to the ecology and evolution of thermal spring microbial communities. These papers address aspects of the ecology of thermophiles and discuss how these organisms influence their environment through their physiological activity (Madigan et al.; Rothschild et al.). Thermal habitats can provide insights into unusual physiologies adapted for conditions similar to the early earth's environment. For example, many laminated stromatolite-like structures are found in the Archaean oceans, which are analogous to the structures formed by microbial mats in thermal springs. Findings from studies in these extant environments may aid interpretations of the nature, distribution, and paleoecology of ancient microorganisms (Lowe et al.; Ward et al.).

The final section in the book addresses some of the applications and potential uses of thermophiles in industry, including the use of carotenoids as antioxidants in food and feed preparations, bioprocessing such as TNT degradation, and coal solubilization and desulfurization (Combie). Given the recent surge of interest in the biotechnological potential of thermophilic microorganisms, there is a need to clarify resource management policies so that microbial resources can be managed effectively to the benefit of all. The volume concludes with a discussion of ways in which the microbial organisms can be managed in areas such as national parks. Some of the specific issues regarding the management of Yellowstone's microbial resources are: inventory and monitoring of the resources, generation and maintenance of research support, habitat protection, legal and ecological ramifications of bioprospecting, and education (Varley et al.). These issues are of global concern and must eventually be confronted by all nations. Many of the initial discussions that started at the meeting in Yellowstone in 1995 were directed toward addressing these issues. The final outcome of those discussions is reflected in the proposed agreement between the biotechnological company, Diversa, and Yellowstone National Park.

The editors thank the authors for their patience in making this project happen. It could not have happened without the help and guidance of Michael Hennelley at Kluwer Academic/Plenum Publishers. Many thanks.

Anna-Louise Reysenbach
Mary Voytek
Rocco Mancinelli

Contents

Thermophiles

Biodiversity, Ecology, and Evolution

1

The Origins of Research on Thermophiles

Thomas D. Brock

---

1. INTRODUCTION

Thermophiles are defined as organisms that can reproduce at high temperatures. How high a temperature? This depends on the group of organisms under consideration. Table 1 illustrates the upper temperature limits for various groups of organisms, as determined by observations in natural habitats. The upper temperature limits for multicellular organisms are much lower than those for unicellular organisms, and the upper limits for eukaryotes are much lower than for prokaryotes. Among the prokaryotes, Archaea have higher upper limits than bacteria (a few Archaea can live at temperatures above that of boiling water). However, within each major group of organisms, only a few members can live at the upper temperature limit.

We call those few species of each group that can live close to the upper temperature limit the thermophiles of that group. Thus, thermophilic fungi and algae (eukaryotes) are those that can live at temperatures above 45–50°C, whereas that would be a rather low temperature for a thermophilic bacterium.

I have defined the thermophilic boundary for prokaryotes as 55 to 60°C (Brock, 1986). Although somewhat arbitrary, this boundary has an ecological and evolutionary basis. Temperatures lower than 50°C are widespread on earth, whereas temperatures higher than 55 to 60°C are much rarer in nature and are associated almost exclusively with geothermal habitats. Thermophilic bacteria have probably evolved in such geothermal habitats; these habitats are the principal focus of this book.

Knowledge of thermophilic bacteria arose from two quite disparate types of research. The first was conventional bacteriological culture procedures, which developed primarily in the field of canning bacteriology. However, because of the applied nature of this research,

---

Thomas D. Brock • University of Wisconsin–Madison, Madison, Wisconsin 53706.

Thermophiles: Biodiversity, Ecology, and Evolution, edited by Reysenbach *et al.* Kluwer Academic / Plenum Publishers, New York, 2001.

Table 1
Upper Temperature Limits for Growth of Various Groups of Organisms[a]

Group	Temperature °C
Animals	
Fish	38
Insects	45–50
Ostracods (crustaceans)	49–50
Plants	
Vascular plants	45
Mosses	50
Eukaryotic microorganisms	
Protozoa	56
Algae	55–60
Fungi	60–62
Prokaryotes	
Bacteria	
Cyanobacteria	70–73
Nonoxygenic phototrophic bacteria	70–73
Heterotrophic bacteria	90
Archaea	
Methane-producing bacteria	110
Sulfur-dependent bacteria	115

[a]Based on Brock (1994).

and the fact that it focused not on growth but on the survival of bacteria at high temperatures, it did not result in the discovery of hyperthermophiles. Standard methods employed by food and canning bacteriologists use culture temperatures around 55°C.

The second type of research on thermophiles involved ecological studies of organisms that live in natural geothermal habitats. However, because of the attractive visible colors of many of the organisms found in hot springs, the focus of such ecological studies was on phototrophs, most of which rarely grow above 60–65°C. Thus, despite many years of research on the microbes of Yellowstone's hot springs, when I first began research in 1965 there was no suggestion that living organisms might be found in boiling water. It was only by careful ecological observations that the existence of these organisms was discovered.

2. EARLY BACTERIOLOGICAL RESEARCH ON THERMOPHILES

Almost at the dawn of bacteriology, the existence of thermophile bacteria was known. Space does not permit more than a cursory glance at the vast research literature in this field (reviews of this early work can be found in Allen (1953), Farrell and Campbell (1969), Farrell and Rose (1967), Fields (1970), and Ingraham (1962), among many others). In his studies on the bacteriology of the Seine River, Miquel (1881) isolated bacteria that grew at 60–70°C, with an upper temperature limit of 75°C. "It is curious," he wrote, "to see a living organism growing in a liquid medium where the hand is harshly burnt in a few seconds." [author's translation]

An extensive review of the literature on bacteria that live at high temperatures was published by Miehe (1907). The upper temperature limit for growth was stated as 75°C, but a few organisms such as *Bacillus calfactor* could remain alive briefly at temperatures as high as 80°C.

Thus, in these early years, bacteriologists knew of the existence of extremely thermophilic bacteria. Subsequently, this knowledge was ignored. Why? By the 1920s the study of thermophiles had become almost the exclusive domain of food bacteriologists, and these high temperature bacteria were forgotten. The temperature used in the canning industry for culturing thermophilic bacteria was established at 55°C, and growth at this temperature then defined a thermophile. Another consequence of the dominance of the food bacteriologist was that the study of the ecology of thermophiles in their natural environments did not take place.

The 55°C culture temperature was so pervasive that by the late 1950s, it became the standard textbook definition of a thermophile. "By incubating enrichment media at very high temperatures (e.g., 55 or 60°C), cultures of thermophilic bacteria can be obtained." (Stanier et al., 1957.) It is hard to imagine today that 60°C was once considered a very high temperature.

3. ECOLOGICAL OBSERVATIONS OF GEOTHERMAL ENVIRONMENTS

Hot springs are found in a wide variety of locations throughout the world, and many biologists, especially after the mid-nineteenth century, made observations of organisms that live in thermal waters (reviewed in Brock, 1967). Examples include Pliny the Elder of the Roman era, the botanist Ferdinand Cohn in 1862, and the biochemist Hoppe-Seyler in 1875. The German phycologist Schwabe (1936) also published an extensive analysis of Icelandic cyanobacteria. There is also an extensive biological literature (Vouk, 1950) that derives from balneology (the uses of geothermal waters for curative purposes).

4. YELLOWSTONE NATIONAL PARK

Extensive observations have been made in Yellowstone National Park over many years of the organisms that live in thermal habitats (reviewed by Brock, 1972, and Brock, 1978). Yellowstone has the highest concentration of thermal features in the world, and its phototrophic mats are attractive and dramatic, so that they fascinated scientists since the first scientific study of the region in the late nineteenth century.

The geologist, W. H. Weed, was the first scientist to study the biology of the Yellowstone springs. His principal interest was in the role organisms might play in the deposition of silica and travertine (Weed, 1888, 1889). Weed noted that the color of the outflow channel was not always due to the mineralogy of the underlying deposit. He recognized that even deposits which were red or yellow could be due to living organisms, which he called algae. Now, we know that these colors are not due to algae but either to cyanobacteria or to nonoxygenic phototrophic bacteria.

After Weed, many scientists studied the microbial mats (reviewed in Brock, 1972), but the most careful and useful observations were those of W. A. Setchell, which unfortunately were never published in full. In his brief but important paper entitled "The Upper Temperature Limit for Life" (Setchell, 1903), he made a clear distinction between phototrophic life and that of nonphototrophic bacteria (called *Schizomycetes* in his day). Setchell showed that the upper temperature limit for phototrophic life was around 75°C, a limit well confirmed by many subsequent observations.[1] He also showed that nonphototrophic bacteria were found at much higher temperatures: "The chlorophylless *Schizophyceae* (or bacterial forms) endure the highest temperatures observed for living organisms, being abundant at 70–71°C and being found in some considerable quantity at 82°C and at 89°C. The temperature of 89°C is the highest at which I have been able to find any organisms living." (Note that Setchell's work in Yellowstone did not include examination of high-temperature habitats microscopically, but involved description only of macroscopic accumulations of microbes.)

Although we do not know where Setchell found organisms living at 89°C, it could well have been at Octopus Spring in the White Creek area, where pink bacteria are easily visible at these temperatures (see Figure 3.2 of Brock, 1978 and Reysenbach et al., 1994).

Unfortunately, studies in Yellowstone during the next 60 years after Setchell's work focused primarily on the taxonomy of cyanobacteria found in the microbial mats (Copeland, 1936; Nash, 1938).[2] Because cyanobacterial taxonomy itself was virtually a meaningless occupation in those years,[3] most of this work is of little value. In fact, many of the organisms which were thought of as cyanophyceae are now known as nonoxygenic phototrophs (Brock, 1968), such as *Chloroflexus* (Pierson and Castenholz, 1974).

5. *THERMUS AQUATICUS*

Although not a hyperthermophile, *Thermus aquaticus* is of interest because it is the source of *Taq* polymerase, the enzyme that plays such a crucial role in the polymerase chain reaction (PCR). In this short section, I review briefly the history of the discovery of this organism.

Thermus aquaticus was first cultured by Hudson Freeze in my laboratory at Indiana University while attempting to culture the pink bacterium of Octopus Spring. Freeze, an honors undergraduate at the time, had spent the summer of 1966 in my laboratory at Yellowstone, and was looking for a research problem to do upon return to the university. He and I took samples of the pink bacteria from Octopus Spring, and we also took samples of cyanobacterial mats from near the upper temperature limit in Mushroom Spring, a nearby hot pool. We set up simple enrichment culture procedures, using a synthetic salt medium that had been developed by R. W. Castenholz for cultivating thermophilic cyanobacteria.

[1]Note that although 73–75°C is the upper temperature limit for phototrophs in Yellowstone, this is not the limiting temperature in other parts of the world. The upper temperature limit for phototrophs is considerably lower in Iceland and New Zealand hot springs, as has been discussed by Castenholz (1969).

[2]Nash was a student of Josephine Tilden, who had done taxonomic research herself in Yellowstone in the late nineteenth and early twentieth centuries (reviewed in Brock, 1972).

[3]The microscopes available were unsuitable for characterizing cyanobacteria. Also, culture methods, so essential for adequate taxonomy, had not yet been developed for cyanobacteria.

To this salt medium were added small amounts of several organic constituents, and incubation was in a water bath at 70°C.

The pink bacteria yielded no positive cultures, but the sample from Mushroom Spring yielded a culture that grew quite well at 70°C and had an upper limit of 79°C. The organism was easily isolated by streaking on agar plates of the same medium and picking colonies. The yellow-pigmented organism was quite characteristic under the microscope, and was clearly a new type. We spent several years characterizing this organism, isolating many more strains from around the world, and studying its biochemistry. The name *T. aquaticus* was chosen after I had selected and rejected several earlier names (Brock, 1995). Strains were deposited in the American Type Culture Collection in early 1969 (Brock and Freeze, 1969), and the organism became available to the scientific community from that source. Many biochemists and industrial scientists obtained cultures, either from me or from the ATCC, and carried out numerous studies. By 1990, a literature survey showed that more than 1000 papers had been published on *T. aquaticus*, mostly since the discovery of *Taq* polymerase (Brock, 1995).

6. DISCOVERY OF EXTREME THERMOPHILES

As research in thermophiles progressed during the past 30 years, ideas of what constitutes an extreme thermophile (hyperthermophile) changed. *Thermus aquaticus* was first characterized as an extreme thermophile (Brock and Freeze, 1969), but because its upper temperature limit is 79°C, it would hardly be called "extreme" today. A simple definition of an extreme thermophile might be one that can live in boiling water. Because water boils in Yellowstone at 92.5°C, an even simpler definition of an extreme thermophile might be one that can live at a temperature higher than 90°C. Indeed, a 90°C boundary for an extreme thermophile is somewhat implied in the paper by Brock et al. (1971) on the bacteria that live in Boulder Spring.

I have reviewed the history of my discovery of extreme thermophiles (Brock, 1978; Brock, 1995; Brock, 1997). This section presents only a short summary.

My original focus in Yellowstone was on the ecology of phototrophs that live at the highest temperatures, and as we have seen, no phototrophs live at temperatures above about 73°C. However, at temperatures above those at which phototrophs lived, I discovered chlorophyll-free organisms in certain hot spring effluents, organisms that were growing so well that they were visible to the naked eye as masses of filamentous streamers.[4] The pink bacteria of Octopus Spring (that live at 85–88°C) are the best example (Reysenbach et al., 1994), but similar populations exist in a number of other effluents.

In 1967, I began to use a sensitive immersion slide technique that demonstrated that bacteria were thriving in virtually all of the boiling pools of neutral to alkaline pH examined in Yellowstone, at temperatures of 92–93°C (the boiling point of water at that altitude). Studies on temperature optima and growth rates showed that these organisms were growing

[4]As noted before, Setchell had earlier described similar organisms. In 1965 when I did my first work, I was unaware of Setchell's paper and had been basing my ideas on an erroneous paper by Kempner (1963).

well and that they were optimally adapted to the temperatures where they were found (Bott and Brock, 1969; Brock et al., 1971).

Similar bacteria were found in boiling springs in New Zealand, where because of the lower altitude, the temperatures are higher (100–101°C) (Brock and Brock, 1971.)

This work showed clearly that bacteria could thrive in boiling water. However, culture work at that time was unsuccessful.

7. *THERMOPLASMA*, *SULFOLOBUS*, AND THE ARCHAEA

In the late 1960s, my interest turned to the highly acidic Yellowstone hot springs, and to other acidic thermal environments such as self-heating coal refuse piles. Culture work from the latter led to the discovery of *Thermoplasma*, a bacterium devoid of a cell wall that could grow at 55°C and pH 2 (Darland et al., 1970). Microscopic and culture studies in sulfur-rich acidic hot springs in Yellowstone led to the discovery of *Sulfolobus*, which resembled *Thermoplasma* in some ways but could grow at higher temperatures both autotrophically and heterotrophically (Brock et al., 1972). The discovery of *Sulfolobus* eventually led to a focus on thermophiles that can metabolize elemental sulfur.

Parallel to my work on the acidophilic thermophiles, Carl Woese was developing his revolutionary techniques for phylogenetic research using 16S RNA. This led to the concept of the archaeabacteria (now called Archaea) (Woese, 1987, 1992).

Originally, only methanogenic bacteria were considered archeabacteria, but about 1980 Woese discovered that *Thermoplasma* and *Sulfolobus* were also Archaea. Shortly thereafter, Wolfgang Zillig, Karl Stetter, and their colleagues began detailed studies of thermophilic Archaea and developed techniques for culturing organisms that grow at temperatures at and above 100°C (Stetter, 1995). The discovery that extreme thermophiles were Archaea led to a vast increase in research on such organisms, because many workers became interested in the evolutionary significance of Archaea. Although very few Archaea have been cultured from Yellowstone sources, phylogenetic studies by the research groups of David Ward and Norman Pace on natural samples have shown that Archaea are widespread in Yellowstone.

8. YELLOWSTONE RESEARCH AND THE DEEP-SEA THERMAL

For many years the Yellowstone work had seemed somewhat exotic to many microbiologists. Although geothermal areas were plentiful throughout the world, they were insignificant on a global scale, and it was not clear how relevant research in Yellowstone might be for broader questions. This attitude changed after the discovery of the deep-sea thermal vents in the late 1970s. Here were extensive habitats with extremely high temperatures, associated with diverse and flourishing life forms (Humphris et al., 1995; Prieur et al., this volume). Because of the Yellowstone work, hypotheses regarding life at high temperature were legitimized. Also, because the vents provided habitats at very high temperatures, it was possible to ask more precisely about the upper temperature for life on earth. (The Archaea in Table 1 that live at temperatures well above the boiling point came from deep-sea vent habitats.)

The Yellowstone work also had another influence on studies of thermal vents. The techniques that had been used to show that bacteria were living and thriving in boiling water could now be applied to the deep-sea vent habitats. It was not just enough to grab samples and culture them. It was essential to show that organisms were (or were not) growing *in situ*. Despite the difficulty of carrying out such experiments in the deep sea, approaches to such work are being carried out and will undoubtedly yield new and exciting information.

Further, because research in Yellowstone is much easier and cheaper than research in the deep sea, new techniques and experiments can be tested there before being taken to the ocean depths. Thus, even scientists interested primarily in thermal vents find a visit to Yellowstone profitable.

9. MICROBIAL PROSPECTING IN THERMAL HABITATS

The discovery of *Taq* polymerase from a strain of *T. aquaticus* isolated from Yellowstone has galvanized the biotechnology industry, and now large numbers of applied researchers visit the park. There is obviously a vast diversity of bacteria in Yellowstone hot springs, and most of these have never been cultured. A few of them may prove of practical use (Madigan and Marrs, 1997; Combie and Runnion, this volume; Varley et al., this volume).

Although there is nothing magic about Yellowstone hot springs, it is easier to sample thermal habitats in Yellowstone than anywhere else in the world. Because the area is completely protected, it is available for long-term research. Even so, there is no guarantee that useful organisms will simply fall out of samples. Clever ecological observations are needed to recognize appropriate sites, and clever culture procedures are needed to enrich organisms of interest. A quick dash in and out of Yellowstone is unlikely to yield anything exciting.

10. CONSERVATION OF YELLOWSTONE'S THERMAL RESOURCES

The Old Faithful symposium of September 1995 upon which this book is based showed how many research groups are currently working in or interested in working in Yellowstone National Park. After the meeting, a number of members of the Geyser Observers Study Association (GOSA), a national organization, pointed out to me that scientists who study in Yellowstone have a responsibility for maintaining the integrity of the habitats they are studying. GOSA members consider that sampling and manipulation of Yellowstone hot springs has the potential for destroying these rather fragile structures. GOSA members feel that some of the manipulative studies that I did on Yellowstone springs, such as adding large amounts of sodium chloride, or altering the routes of the outflow channels exhibits an arrogance that they consider unacceptable. Although I was always cautious in my own work to avoid any irreversible change in the thermal feature I studied, the GOSA members have pointed out that what I consider harmless may be a matter of opinion. Clearly, there is room for debate on these issues and to call attention to these concerns, I agreed to insert this brief section into this chapter.

Fortunately, microbiologists require only very small samples for most of their work, so that a conservation ethic is not incompatible with thorough research studies.

11. SUMMARY

Yellowstone National Park and other geothermal areas of the world have yielded a large number of interesting hyperthermophilic bacteria. Such organisms may have considerable biotechnological potential, but perhaps of greatest interest are the clues they give us about the origin and evolution of life.

The discovery of hyperthermophiles in Yellowstone and elsewhere came about because of careful ecological studies in the geothermal habitats themselves. It was only by such ecological studies that the 55°C dogma inherited from food bacteriologists could be abandoned. Once it was proved by ecological studies that bacteria lived and thrived in boiling water, extensive culture work could be justified.

Unfortunately, the ecological aspects of microbial research in Yellowstone have not found great application in the years since I terminated my research. Only a few laboratories have seen fit to make *in situ* experiments. Although it was through pure basic research in which *T. aquaticus* and hyperthermophiles were discovered (Brock, 1997), most research today seems to be on applied aspects (Madigan and Marrs, 1997). There are many fascinating basic research studies yet to be carried out in Yellowstone, and the Yellowstone geothermal habitats remain most favorable for microbial ecological research.

REFERENCES

Allen, M. B. 1953. The thermophilic aerobic sporeforming bacteria. *Bacteriol. Rev.* **17:**125–173.

Bott, T. L., and Brock, T. D. 1969. Bacterial growth rates above 90°C in Yellowstone hot springs. *Science* **164:**1411–1412.

Brock, T. D. 1967. Life at high temperatures. *Science* **158:**1012–1019.

Brock, T. D. 1968. Taxonomic confusion concerning certain filamentous blue-green algae. *J. Phycol.* **4:**178–179.

Brock, T. D. 1972. One hundred years of algal research in Yellowstone National Park. In Desikachary, T. V. (ed.), *Taxonomy and biology of blue-green algae* (pp. 393–405). Madras, India: Center for Advanced Study in Botany.

Brock, T. D. 1978. *Thermophilic microorganisms and life at high temperatures.* New York: Springer-Verlag.

Brock, T. D. 1986. Introduction: An overview of the thermophiles. In Brock, T. D. (ed.) *Thermophiles: General, molecular, and applied microbiology* (pp. 1–16). New York: Wiley.

Brock, T. D. 1994. *Life at high temperatures.* Yellowstone National Park, WY: Yellowstone Association.

Brock, T. D. 1995. The road to Yellowstone—and beyond. *Annu. Rev. Microbiol.* **49:**1–28.

Brock, T. D. 1997. The value of basic research: Discovery of *Thermus aquaticus* and other extreme thermophiles. *Genetics* **146:**1207–1210.

Brock, T. D., and Brock, M. L. 1971. Microbiological studies of thermal habitats of the central volcanic region, North Island, New Zealand. *N.Z. J. Mar. Freshwater Res.* **5:**233–257.

Brock, T. D., and Freeze, H. 1969. *Thermus aquaticus* gen. n. and sp. n., a nonsporulating extreme thermophile. *J. Bacteriol.* **86:**708–712.

Brock, T. D., Brock, M. L., Bott, T. L., and Edwards, M. R. 1971. Microbial life at 90°C: The sulfur bacteria of Boulder Spring. *J. Bacteriol.* **107:**303–314.

Brock, T. D., Brock, K. M., Belly, R. T., and Weiss, R. L. 1972. *Sulfolobus*: A new genus of sulfur-oxidizing bacteria living at low pH and high temperature. *Archiv fur Mikrobiologie* **84:**54–68.

Castenholz, R. W. 1969. The thermophile cyanophytes of Iceland and the upper temperature limit. *J. Phycol.* **5:**360–368.

Copeland, J. J. 1936. Yellowstone thermal Myxophyceae. *Ann. N.Y. Acad. Sci.* **36:**1–229.

Darland, G., Brock, T. D., Samsonoff, W., and Conti, S. F. 1970. A thermophilic, acidophilic mycoplasma isolated from a coal refuse pile. *Science* **170:**1416–1418.

Farrell, J., and Campbell, L. L. 1969. Thermophilic bacteria and bacteriophages. *Adv. Microb. Physiol.* **3:**83–109.

Farrell, J., and Rose, A. H. 1967. Temperature effects on microorganisms. In Rose, A. H. (ed.), *Thermobiology* (pp. 147–218). London: Academic Press.

Fields, M. L. (1970). The flat sour bacteria. *Adv. Food Res.* **18**:163–217.

Humphris, S. E., Zierenberg, R. A., Mullineaux, L. S. and Thomson, R. E. (eds.). 1995. *Seafloor hydrothermal systems—physical, biological, and geological interactions.* Washington, D.C.: American Geophysical Union.

Ingraham, J. 1962. Temperature relationships. In Gunsalus, I. C., and Stanier, R. Y. (eds.), *The bacteria*, Vol. IV (pp. 265–296). New York: Academic Press.

Kempner, E. 1963. Upper temperature for life. *Science* **142:**1318–1319.

Madigan, M. T. and Marrs, B. L. 1997. Extremophiles. *Sci. Am.* **276:**82–87.

Miehe, H. 1907. *Die Selbsterhitzung des Heus.* Jena, Germany: Verlag von Gustav Fischer.

Miquel, M. Bulletin de la statistique municipale de la ville de Paris, December 1879. *Annuaire de l'observatoire de Montsouris pour 1881.* p. 464.

Nash, A. 1938. The cyanophyceae of the thermal regions of Yellowstone National Park, U.S.A., and of Rotorua and Whakarewarewa, New Zealand, with some ecological data. Ph.D. Thesis, University of Minnesota, Minneapolis.

Pierson, B. K., and Castenholz, R. W. 1974. A phototrophic gliding filamentous bacterium of hot springs, *Chloroflexus aurantiacus* gen. and sp. nov. *Arch. Microbiol.* **100:**5–24.

Reysenbach, A.-L., Wickham, G. S., and Pace, N. R. 1994. Phylogenetic analysis of the hyperthermophilic pink filament community in Octopus Spring, Yellowstone National Park. *Appl. Environ. Microbiol.* **60:**2113–2119.

Schwabe, G. H. 1936. Beiträge zur Kenntnis isländischer Thermalbiotope. *Archiv für Hydrobiologie* **6**(Supplement):161–352.

Setchell, W. A. 1903. The upper temperature limit for life. *Science* **17:**934–937.

Stanier, R. Y., Doudoroff, M., and Adelberg, E. A. 1957. *The microbial world.* Englewood Cliffs, NJ: Prentice-Hall.

Stetter, K. O. 1995. Microbial life in hyperthermal environments. *ASM News* **61:**285–290.

Vouk, V. 1950. *Grundriss zu einer Balneobiologie der Thermen.* Basel, Switzerland: Birkhäuser Verlag.

Weed, W. H. 1888. Formation of travertine and siliceous sinter by the vegetation of hot springs. *Rep. U.S. Geol. Surv.* **9:**619–676.

Weed, W. H. 1889. The vegetation of hot springs. *Am. Naturalist* **23:**394–400.

Woese, C. 1987. Bacterial evolution. *Microbiol. Rev.* **51:**221–271.

Woese, C. 1992. Prokaryotic systematics: The evolution of a science. In Balows, A., Truper, H., Dworkin, M., Harder, W., and Schleifer, K.-H. (eds.), *The prokaryotes II*, Vol. 1, (pp. 3–18). New York: Springer-Verlag.

2

Deep-Sea Thermophilic Prokaryotes

Daniel Prieur, Mary Voytek, Christian Jeanthon, and Anna-Louise Reysenbach

1. INTRODUCTION

One of the major biological discoveries of the last few decades is the rich and unusual micro- and macrofauna clustered around deep-sea hydrothermal vents. Since their discovery in 1977, deep-sea hydrothermal vents have been explored and sampled by biologists interested in the ecology of extreme environments. However, due to the geological complexity and the many methodological problems associated with their remoteness (such as the necessity for manned submersibles or remotely operated vehicles), deep-sea vent systems are difficult to sample and therefore microbial studies have been relatively limited.

Several reviews dealing with the taxonomy and the terrestrial and marine ecology of thermophiles have been published recently (Prieur, 1992; Prieur et al., 1995; Bloch et al., 1995; Stetter, 1996, 1998). In this chapter, we provide an overview of thermophiles and hyperthermophiles from deep-sea hydrothermal vents, relying both on isolates and the metabolic information derived from autoecological studies and on phylogenetic studies examining the diversity of bacteria and Archaea. Whenever possible, we also compare hydrothermal vent communities and their terrestrial counterparts, the thermal springs of Yellowstone National Park.

Daniel Prieur and Christian Jeanthon • CNRS, UPR 9042 and UPMC, Station Biologique, Roscoff and Université de Bretagne Occidentale, Brest, France. **Mary Voytek** • United States Geological Society, MS430, Reston, Virginia 20192. **Anna-Louise Reysenbach** • Portland State University, Department of Environmental Biology, Portland, Oregon 97201.

Thermophiles: Biodiversity, Ecology, and Evolution, edited by Reysenbach *et al.* Kluwer Academic/Plenum Publishers, New York, 2001.

2. HYDROTHERMAL VENT ENVIRONMENTS

The ocean floor is continually being renewed at seafloor spreading centers. In these tectonically active areas, seawater penetrates through cracks in the ocean floor and is then chemically altered by the interaction of heat and the surrounding rock within the crust. The heated, chemically modified low-density fluid is forced back to the ocean floor by convection. The fluid is hot (250–400°C and remains liquid because of the hydrostatic pressure, about 26 Mpa at 2600 m), acidic, and enriched in metals and reduced compounds, such as iron, methane, and hydrogen sulfide. Chemolithotrophic microbial communities that are considered the primary producers in these systems can use these compounds as energy sources. When the hot, reduced fluid mixes with cold oxygenated deep ocean water, minerals precipitate and construct spectacular sulfidic mineral structures ("chimneys") from which the hot fluid vents. Hydrothermal vent fields (i.e., single sites of hydrothermal emissions) are generally tens of meters in diameter and may be interconnected via subterranean conduits. Most of the known hydrothermal vents in the ocean are located along the global mid-ocean ridge system, back-arc basins, and hot spots in the deep sea and can be found at depths greater than 3500 m and as shallow as 120 m.

3. BIOLOGICAL COMMUNITIES

Deep-sea hydrothermal vents are like oases on the seafloor and they provide numerous habitats and chemical energy for the growth of macro- and microorganisms. The rich and visually spectacular macrobiotic communities at deep-sea hydrothermal vents are restricted to areas where hydrothermal fluids are diluted with cold ambient sea water resulting in a diffuse flow of warm fluids (10–40°C). The microbial communities form thick mats in these macrofaunal communities; they colonize the surfaces of vent animals and exist as animal–bacterial symbioses. They are also closely associated with the high temperature environments such as the subsurface conduits of vents, suspended within the effluent itself or attached to mineral or sedimentary surfaces that surround the vent openings.

Due to the variation and dynamic chemical and physical nature of their biotopes, thermophile microorganisms in both terrestrial and submarine hydrothermal vent environments occupy distinct niches characterized by geochemical gradients. The high temperatures and steep thermal and chemical gradients encountered in these environments determine the types of organisms that occur in the different microhabitats. The ecological consequence of such extreme conditions is the establishment of prokaryotic communities in both systems that share many of the same genera and species and have developed similar metabolic strategies. However, one obvious difference between the two systems is the absence of the visually dominant forms of terrestrial hot springs, the pigmented phototrophic prokaryotes. These phototrophs are replaced at submarine vents to a large extent by both aerobic and anaerobic chemolithotrophs.

4. ECOLOGICAL STUDIES

It is difficult to collect samples in the deep sea, particularly from hydrothermal environments. As a consequence, ecological data on deep-sea thermophilic communities are rare, except for those deduced from laboratory studies of pure strains. Much of what we know about the physiology of prokaryotes from hydrothermal vents is either inferred from chemistry or based on studies performed on isolates.

4.1. Locating the Niche

One of the most basic and difficult questions to address regarding the free-living microbial communities at deep-sea vents is to locate the precise ecological niches of the microorganisms. Many of the strains reported in the literature have been isolated from pieces of hydrothermal chimneys. A few strains have been isolated from fluid samples (Erauso et al., 1993). However, it is well known that hydrothermal fluid samples frequently contain small chimney debris and material flushed as exiting water scours the vent walls. Another strain has been isolated from tissues of an alvinellid worm that builds its tube on the outer wall of hydrothermal chimneys (Pledger and Baross, 1989). Assigning specific organisms to a particular biotope becomes difficult in a dynamic system where material is naturally being transported and sampling procedures can also be disruptive. The best studies have used *in situ* techniques, such as microelectrodes, or *in situ* fluorescent oligonucleotide hybridization of horizons of chimney samples (Harmsen et al., 1997a) or lipid biomarker analysis of chimney horizons (Hedrick et al., 1992).

4.2. Microbial Abundances

The distribution and abundance of hydrothermal microorganisms are fundamental, yet poorly addressed ecological parameters. Prokaryotes associated with vent communities can be free living bacteria within hydrothermal fluids or can colonize vent surfaces or be found in animal–bacterial symbioses. Several techniques have been used to estimate the abundances of microorganisms in these biotopes, including Most Popular Number (MPN) determination, acridine orange or 4′,6′-diamidino-2-phenylindole (DAPI) direct counts, lipid analysis, or more recently fluorescent nucleic acid probes (Harmsen et al., 1997a,b). Several estimates have been reported for microorganisms within chimney walls; numbers fluctuate from 10^4 to 10^9 cells per gram of chimney material (Harmsen et al., 1997a). The outer horizons of chimneys have the highest numbers and are dominated by Bacteria. The low numbers associated with the internal portions of the chimneys are almost exclusively attributed to Archaea. In many cases, the inner portions of chimneys may be exposed to fluids hotter than 350°C, which raises the question of the exact habitat of these microorganisms and the lack of good sampling stategies to locate microniches. Estimates for microbial numbers in hydrothermal fluids range from 10^5 to 10^9/mL. In many cases, the densities of microorganisms associated with vent fields are very high, such that the seawater appears milky or the seafloor is carpeted by white and yellow microbial mats.

4.3. Origin and Biogeography

The ecology of deep-sea thermophiles is still understudied, but there have been several observations that address the biogeography of these organisms. Several species such as *Thermococcus litoralis* have been isolated from a variety of disparate environments: coastal and deep hydrothermal areas (Neuner et al., 1990), the volcanic plume following a submarine eruption (Huber et al., 1990) and from offshore (Stetter et al., 1993) and continental oil fields (L'Haridon et al., 1995). In the latter cases, the oil reservoirs may be inoculated with thermophiles in the seawater. This possible explanation is supported by observations that thermophiles have been isolated from cold seawater samples far removed from hydrothermal areas (Stetter, 1998) and certain strict anaerobes can tolerate oxygen at low temperatures (Huber et al., 1990; Marteinsson et al., 1997). However, this hypothesis does not explain the occurrence of the same microorganisms in continental reservoirs that are not in direct contact with the sea (L'Haridon et al., 1995). In this case, it is possible that these microorganisms are the progeny of microbes that were trapped in the sediment that ultimately gave rise to the oil reservoir or they have been introduced into the reservoirs during commercial exploitation. In the latter case, the "contaminating" inocula would come from surface water or deep continental aquifers.

4.4. Barotolerance and Barophily

Hydrostatic pressure is one potential environmental constraint experienced by submarine thermophiles and not by their terrestrial counterparts. The foundations of pressure effects on microbial activity were established by studies on deep-sea psychrophiles (see Deming and Baross, 1993, for review). Three basic types of pressure responses have been documented: barotolerance, barosensitivity, and barophily. One interesting effect of pressure on some strains is that the optimum and maximal growth temperatures are shifted up by a few degrees compared to optimal and maximal growth temperature under atmospheric pressure. These strains are referred to as obligate barophiles, but the term barodependent is probably more correct. For some barophilic thermophiles, the optimal pressure for growth appeared higher than the pressure at the depth of collection (Deming and Baross, 1993). From these observations, Deming and Baross suggested that thermophiles could inhabit subterranean reservoirs where they are exposed to both hydrostatic and lithostatic pressure and hydrothermal black smokers can then serve as "windows" into these reservoirs, extending thermophilic habitats considerably.

4.5. Temperature: Optima and Limits

Prokaryotes have been detected and isolated from samples of hydrothermal vent fluids at temperatures exceeding 350°C, but thermophiles and hyperthermophiles isolated from shallow and deep-sea hydrothermal systems have never been observed growing at temperatures above 113°C. The reported growth range for hyperthermophiles is from 60–113°C and the optima from 86–105°C (Stetter, 1982; Blochl et al., 1997). All archaeal isolates thrive at temperatures from 80–113°C and cannot grow at temperatures lower than 60°C.

5. DIVERSITY: THERMOPHILIC AND HYPERTHERMOPHILIC ISOLATES

Until relatively recently, descriptions of microbial diversity were limited to what could be grown in the laboratory. However, with the development of molecular phylogenetic approaches for studying microbial diversity, the paucity of information on natural microbial diversity has been appreciated more. These culture-independent approaches provide little or no information on the physiological attributes of the organisms themselves.

Therefore, descriptions of microbial diversity are best approached using both culture enrichment techniques and culture-independent molecular methods.

Examples of described thermophilic and hyperthermophilic prokaryotes isolated from deep-sea samples are given in Table 1. Most of the hyperthermophilic isolates from deep-sea hydrothermal vents belong to the domain *Archaea*, except isolates of the deeply rooted lineages within the *Bacteria*, the *Thermotogales* and a novel recently described lineage, the type strain, *Desulfurobacterium thermolithotrophum*. Deep-sea hydrothermal hyperthermophilic *Archaea* include members of the genera *Pyrodictium*, *Pyrococcus*, *Archaeoglobus*, *Desulfurococcus*, and *Thermococcus*. No mesophilic or psychrophilic representatives from the marine archaeal groups I and II (DeLong, 1992) have been isolated from vent systems. However, Moyer et al. (1998) found crenarchaeal 16S rRNA gene sequences associated with microbial mats at Loihi Seamount that are closely related to the low temperature crenarchaea found in nonthermophilic habitats (see below).

The high temperature, deep-sea hydrothermal fluids have a relatively low pH (3.7–4.8) (Von Damm, 1995). It is surprising that thermoacidophiles have not yet been isolated from deep-sea vents. Thermoacidophilic organisms such as *Sulfolobus* are common inhabitants of terrestrial hot springs (Brock, 1978). Despite many attempts to cultivate hyperthermophilic aerobic acidophiles such as *Sulfolobus* from deep-sea vents, they have never been isolated from deep-sea vent systems. It is possible that these organisms cannot tolerate the steep chemical fluctuations that exist at deep-sea vents as the anoxic hot, acidic hydrothermal fluids mix with the oxic cool, alkaline bottom seawater. It has also been suggested that *Sulfolobus* is unable to grow in saline conditions (Stetter, 1996). Furthermore, a member of the *Sulfolobales*, *Sulfurococcus mirabilis*, cannot maintain its internal pH at 4°C. Additionally, the acid deep-sea hydrothermal fluids mix rapidly with seawater, raising the pH, and niches (low pH, high temperatures) favorable for thermoacidophiles may not yet have been detected. Besides the *Sulfolobales* exception, the other main archaeal terrestrial geothermal orders have abyssal relatives. *Thermoproteales* are represented by heterotrophic and sulfur-respiring deep-sea *Desulfurococcus* strains and by the shallow marine isolate *Staphylothermus marinus*. Similarily, *Pyrodictiales* are represented by *Pyrodictium abyssi*, a heterotroph whose maximum temperature for growth is 110°C. Although the other nonabyssal species of *Pyrodictium* are autotrophic and utilize hydrogen and elemental sulfur as electron donor and acceptor, respectively, this deep-sea species is heterotrophic. The only sulfate-reducing archaeum, *Archaeoglobus profundus* belongs to the *Archaeoglobales* and is also an obligate heterotroph. Deep-sea thermophilic methanogens are represented by two orders, *Methanococcales* and *Methanopyrales*, and two strict autotrophic species, *Methanococcus jannaschii* (although strains of *M. jannaschii* can also use formate) and *Methanopyrus kandleri*; the latter has also been isolated from shallow geothermal areas.

Most microbiological studies at deep-sea hydrothermal vents have focused on hyperthermophiles, so the list of thermophilic bacteria reported from the deep-sea vents is rather short. Aerobic thermophiles have only been reported very recently (Marteinsson et al., 1995, 1996). Hydrogen-oxidizing bacteria belonging to the *Aquificales* must exist within these biotopes because both electron donor (hydrogen) and acceptor (oxygen) are abundant. Huber et al. (1992) isolated a member (*Aquifex pyrophilus*) from a shallow (106 m) hydrothermal vent near Iceland, however, relatives of this group are yet to be isolated from deep-sea vents. One surprising recent discovery was the isolation of bacterial chemo-

Table 1
Examples of Thermophilic and Hyperthermophilic Bacteria and Archaea from Deep-Sea Hydrothermal Vents

	Examples of sample site	General metabolism	Reference
Bacteria			
Spore-forming gram+ (*Bacillus*-like)	Mid-Atlantic Ridge (23°N, 44°W) Lau Basin (22°S, 176°E) Guaymas Basin 27°N, 112°W	Heterotroph, fermentative	Marteinsson et al., 1996
Non-spore-forming (*Thermus*-like)	Mid-Atlantic Ridge (23°N, 44°W) Guaymas Basin 27°N, 112°W	Heterotroph	Marteinsson et al., 1996
Desulfurobacterium	Widespread, associated with sulfide deposits	Sulfur-reducing chemolithotroph	L'Haridon et al., 1998
Archaea			
Desulfurococcales			
Desulfurococcus S, SY	EPR 11°N, 104°W	Sulfur-reducing heterotroph	Jannasch et al., 1988
Staphylothermus marinus	EPR 11°N, 104°W	Sulfur-reducing heterotroph	Fiala et al., 1986
Pyrodictiales			
Pyrodictium abyssi	Guaymas Basin 27°N, 112°W	Sulfur reducer	Pley et al., 1991
Archaeoglobales			
Archaeoglobus profundus	Guaymas Basin 27°N, 112°W	Sulfate reducer	Burggraf et al., 1990
Thermococcales			
Thermococcus spp. *Pyrococcus* spp.	Widespread	Sulfur-reducing heterotroph	e.g., Antoine et al., 1995; Erauso et al., 1993; Godfroy et al., 1996; Huber et al., 1995; Jannasch et al., 1992; Kobayashi et al., 1994; Kwak et al., 1995; Marteinsson et al., 1995; Pledger and Baross, 1989; Raguenes et al., 1995
Methanococcales			
Methanococcus jannaschii	EPR 21°N, 109°W	Methanogen	Jones et al., 1983
Methanococcus fervens	Guaymas Basin 27°N, 112°W	Methanogen	Zhao et al., 1988; Jeanthon et al., 1999
Methanopyrales			
Methanopyrus kandleri	Guaymas Basin 27°N, 112°W	Methanogen	

lithotrophic sulfur reducers that form an entire new lineage, branching between the *Aquificales* and *Thermotogales* (L'Haridon et al., 1998).

However, by far most of the novel deep-sea thermophilic prokaryotic isolates described thus far, belong to the *Thermococcales*. Optimal temperatures for growth of the two known genera (*Pyrococcus* and *Thermococcus*) are above 80°C, and doubling times are about 30 minutes. All isolates are heterotrophic, strictly anaerobic that ferment complex proteinaceous substrates, carbohydrates, and polymers such as chitin. Hydrogen is one of the fermentation products and is inhibitory to growth. The *Thermococcales* avoid this inhibition by using sulfur as a sink for the hydrogen and producing hydrogen sulfide.

There are several explanations why the *Thermococcales* are so dominant in deep-sea thermophile collections. Primarily, the cultivation of autotrophs in the laboratory is often more difficult than growing heterotrophs. Additionally, *Thermococcales* are relatively tolerant of the deep-sea sampling conditions. Upon collection, the samples are cooled and oxygenated as they pass through the water column. The *Thermococcales* are extremely sensitive to oxygen at high growth temperatures, but they tolerate aerobic conditions at low temperatures quite well, particularly at low nutrient concentrations (Marteinsson et al., 1997). Furthermore, Martiensson et al. (1995) reported twenty-one isolates that were obtained from a single deep-sea vent smoker, and six of them were novel genomic strains based on DNA-DNA hybridization. Such culturable diversity is not easy to explain because all of the isolates showed very similar phenotypes, particularly with regard to their carbon sources. However, their individual growth temperatures ranges varied. This phenotypic plasticity would be an advantage in hydrothermal habitats where temperatures fluctuate (Chevaldonne et al., 1991). Therefore, the different sulfur-reducing strains growing within a single hydrothermal chimney would constitute a functional community within an environment whose temperature fluctuates.

The reduced taxonomic novelty of deep-sea thermophiles may also be attributed to our inability to enrich and isolate representatives by conventional microbiological methods. Conventional techniques used for isolating thermophiles were designed for shallow vent samples. One major parameter in the deep-sea is hydrostatic pressure that increases by 0.1 MPa per 10 m. Most deep-sea thermophile isolates have been cultivated in the laboratory under atmospheric pressures. However, recently, a novel *Thermococcales* strain (tentatively named *Thermococcus barophilus*) has been isolated under *in situ* pressure and temperature (Marteinsson et al., 1996). The maximal and optimal temperatures for growing this isolate are not modified by pressure, but the doubling time of the organism decreases twofold under hydrostatic pressure. Additionally, this isolate produces different cell proteins when grown under pressure or not.

Many species such as *Staphylothermus marinus* (Fiala et al., 1986), *Methanopyrus kandleri* (Kurr et al., 1991), *Methanococcus igneus* (Jeanthon et al., unpublished data) have been found in both coastal and abyssal hydrothermal areas. Furthermore, some species such as *Thermococcus litoralis* are ubiquitous. It has been isolated from coastal and deep-sea vents (Huber et al., 1990; Bonch-Osmolovskaya, pers. com.) and also from offshore (Stetter et al., 1993) and continental deep oil reservoirs (L'Haridon et al., 1995).

Until recently, no genera specific to the deep sea have been isolated; however, the isolation of the chemolithotrophic sulfur-reducing *Desulfurobacterium* changed this observation, although to our knowledge, isolation of this organism from shallow marine vents has not yet been attempted. With this exception, it is likely that novel genera will continue to be isolated from deep-sea hydrothermal vents.

6. ASSESSMENTS OF MOLECULAR DIVERSITY

It is well recognized that the majority of microorganisms found in nature have not yet been cultivated. Molecular phylogenetic techniques based on the small subunit rRNA (16S rRNA) have provided a mechanism to access information about the diversity of these microbial communities, alleviating the requirement to grow the organisms in the laboratory. Use of these techniques has resulted in the detection of previously unknown microorganisms and their phylogenetic positions have been used to infer information about their potential metabolism and overall microbial interactions. Several 16S rRNA-based techniques have been employed to examine the diversity of vent communities. Using whole cell hybridizations with 16S rRNA-targeted probes targeting the bacterial domain and other specific probes that included most of the thermophilic members of the genus *Bacillus*, most species of the genus *Thermus*, the genera *Thermotoga* and *Thermosipho*, and the *Aquificales* order, Harmsen et al. (1997) observed that the overall diversity of vent communities was much larger than that assessed by using culture-related to related to the 16S rRNA gene sequences from the Yellowstone National Park black filamentous organisms and those reported in Chapter 7 (see Chapter 7).

7. BIOPROSPECTING AND BIOTECHNOLOGY

The contribution of deep-sea thermophiles to biotechnological applications is still modest. Many thermostable enzymes have been identified and studied and several have been purified. DNA polymerases have been isolated from deep-sea thermophiles, and some are already commercially available (e.g., *Thermococcus litoralis*, *Thermatoga maritima*, and *Pyrococcus furiosus*). The purification and development of industrial applications of hydrolases, such as amylases, lipases, and proteases, are well underway for deep-sea thermophiles. Natural product chemists are now looking to vent thermophiles for novel metabolites to be used in developing new pharmaceuticals. Hyperthermophilic *Thermococcus sp.* form organic sulfur compounds similar to lenthionin, some of which are pharmaceutically active (Ritzau et al., 1993). A more recent and exciting contribution to biotechnology pertains to genetic studies of anaerobic hyperthermophiles. To date, research in this area has not been possible because genetic tools such as cloning and expression vectors, which are commonly used to study mesophiles, have not been available. A small (3.5 kb) multicopy plasmid was recently discovered in strain GE5 of the deep-sea species *Pyrococcus abyssi* (Erauso et al., 1996; Aagard et al., 1996). This plasmid may prove to be a first generation vector for use in genetic studies of anaerobic thermophiles. Research is progressing rapidly as an increasing number of researchers become involved, and undoubtedly this field of science will produce major discoveries within the next ten years.

8. HYDROTHERMAL VENTS AND THE ORIGIN OF LIFE

Many have suggested that microbial communities found in terrestrial hot springs may provide information on the evolution of life on earth, and some have suggested that life may have originated at hydrothermal vents (Cairns-Smith et al., 1992). There are several lines of evidence to support this theory. The geochemical disequilibrium that exists in hydrothermal fluids provides a very readily available energy source, and the carbon dioxide is an excellent carbon source for life (Shock, 1996). Others have postulated that charged molecules such as pyrite are good molecules for assembling the first biomolecules of life (Wächterhäuser, 1988). Pyrite synthesis occurs readily at deep-sea hydrothermal vents and could be the surface on which life may have arisen. Additionally, inhabiting the deep sea confers another benefit, the ability to survive cosmological events that occurred early in the earth's history. This meteroritic bombardment would have sterilized the earth's surfaces (Maher and Stevenson, 1988), and only thermophilic organisms deep in the subsurface could have survived these impacts (Stevens, 1997). Additionally, although still highly debated, the rooted phylogenetic tree of life places thermophilic hydrothermal vent isolates closest to the base of the tree and the last common ancestor (LCA). This positioning has been used to support the theory that life began in a high-temperature environment.

9. SUMMARY

Many strains of thermophilic and hyperthermophilic bacteria and *Archaea* have been isolated from deep-sea hydrothermal vents, and several novel species have been described. Except for thermoacidophilic organisms, all of the genera already known from shallow marine and terrestrial hydrothermal environments have been found at deep-sea vents. Indeed, several species seem to have a ubiquitous distribution, and occurring in shallow and deep marine habitats and also in marine and continental oil fields. Although deep-sea hydrothermal vents are exposed to elevated hydrostatic pressure, no obligate barophilic thermophiles have yet been isolated from depth. However, several deep-sea hyperthermophiles have demonstrated a barophilic response when exposed to *in situ* pressure. The limited molecular phylogenetic assessments of the microbial diversity at deep-sea hydrothermal vents do illustrate that we have only started to explore the diverse richness in these unusual ecosystems.

REFERENCES

Aagaard, C., Leviev, I., Aravalli, R. N., Forterre, P., Prieur, D., and Garrett, G. 1996. General vectors for archaeal hyperthermophiles: strategies based on a mobile intron and a plasmid. *FEMS Microbiol. Rev.* **18**:93–104.

Antoine, A., Guezennec, J., Meunier, J-R., Lesongeur, F., and G. Barbier. 1995. Isolation and characterization of extremely thermophilic archaebacteria related to the genus *Thermococcus* from deep-sea hydrothermal Guaymas Basin. *Curr. Microbiol.* **31**:186–192.

Blöchl, E., Burggraf, S., Fiala, G., Lauerer, G., Huber, G., Huber, R., Rachel, R., Segerer, A., Stetter, K. O., and Volk, P. 1995. Isolation, taxonomy and phylogeny of hyperthermophile microorganisms. *World J. Microbiol. Biotech.* **11**:9–16.

Blöchl, E., Rachel, R., Burggraf, S., Hafenbradl, D., Jannasch, H. W., and Stetter, K. O. 1997. *Pyrolobus fumarii*, gen. and sp. nov., represents a novel group of Archaea, extending the upper temperature limit for life to 113°C. *Extremophiles* **1**:14–21.

Brock, T. D., and Freeze, H. 1969. *Thermus aquaticus* gen. n. and sp. n., a non-sporulating extreme thermophile. *J. Bacteriol.* **98**:289–297.

Brock, T. D., Brock, K. M., Belly, R. T., and Weiss, R. L. 1972. *Sulfolobus*: A new genus of sulfur-oxidizing bacteria living at low pH and high temperature. *Arch. Microbiol.* **84**:54–68.

Burggraf, S., Jannasch, H. W., Nicolaus, B., and Stetter, K. O. 1990. *Archaeoglobus profundus* sp. nov., represents a new species within the sulfate-reducing archaeabacteria. *Syst. Appl. Microbiol.* **13:**24–28.

Cairns-Smith, A. G., Hall, A. J., and Russell, M. J. 1992. Mineral theories of the origin of life and an iron sulfide example. *Orig. Life Evol. Biosphere* **22:**161–180.

Chevaldonne, P., Desbruyeres, D., and Le Haitre, M. 1991. Time-series of temperature from three deep-sea hydrothermal vent sites. *Deep-Sea Res.* **38:**1417–1430.

DeLong, E. F. 1992. Archaea in coastal marine environments. *Proc. Natl. Acad. Sci. USA* **89:**5685–5689.

Deming, J. W., and Baross, J. A. 1993. Deep-sea smokers: Windows to a subsurface biosphere. *Geochim. Cosmochim. Acta* **57:**3219–3230.

Donk, P. J. 1920. A highly resistant thermophilic organism. *J. Bacteriol.* **5:**373.

Erauso, G., Reysenbach, A. L., Godfroy, A., Meunier, J. R., Crump, B., Partensky, F., Baross, J. A., Marteinsson, V. T., Barbier, G., Pace, N., and Prieur, D. 1993. *Pyrococcus abyssi* sp. nov., a new hyperthermophilic archaeon isolated from a deep-sea hydrothermal vent. *Arch. Microbiol.* **160:**338–349.

Fiala, G., Stetter, K. O., Jannasch, H. W., Langworthy, T. A., and Madon, J. 1986. *Staphylothermus marinus* sp. nov. represents a novel genus of extremely thermophilic submarine heterotrophic archaebacteria growing up to 98°C. *Syst. Appl. Microbiol.* **8:**106–113.

Godfroy, A., Meunier, J-R., Guezennec, J., Lesongeur, F., Raguenes, G., Rimbault, A., and Barbier, G. 1996. *Thermococcus furnicolans* sp. nov. a new hyperthermophilic Archaeum isolated from deep-sea hydrothermal vent in North Fiji basin. *Int. J. Syst. Bact.* **46:**1113–1119.

Gonzalez, J. M., Kato, C., and Horikoshi, K. 1995. *Thermococcus peptonophilus* sp. nov., a fast growing, extremely thermophilic archaebacterium isolated from deep-sea hydrothermal vents. *Arch. Microbiol.* **164:**159–164.

Harmsen, H. J. M., Jeanthon, C., and Prieur, D. 1996. Whole-cell hybridization with fluorescent probes of cells extracted from deep-sea hydrothermal vent chimneys. *Thermophiles 96*, September 1996, Athens, GA.

Harmsen, H. J. M., Prieur, D., and Jeanthon, C. 1997a. Distribution of microorganisms in deep-sea hydrothermal vent chimneys investigated by whole-cell hybridization and enrichment culture of thermophilic subpopulations. *Appl. Environ. Microbiol.* **63:**2876–2883.

Harmsen, H. J. M., Prieur, D., and Jeanthon, C. 1997b. Group-specific 16S rRNA-targeted oligonucleotide probes to identify thermophil bacteria in marine hydrothermal vents. *Appl. Environ. Microbiol.* **63:**4061–4068.

Hedrick, D. B., Pledger, R. D., White, D. C., and Baross, J. A. 1992. In situ microbial ecology of hydrothermal vent sediments. *FEMS Microb. Ecol.* **101:**1–10.

Huber, R., and Stetter, K. O. 1992. The order *Thermoproteales*. In: *The prokaryotes* (pp. 677–683). Heidelberg: Springer Verlag.

Huber, R., Stoffers, P., Chemine, J. L., Richnow, H. H., and Stetter, K. O. 1990. Hyperthermophilic archaebacteria within the crater and open-sea plume of erupting Macdonald Seamount. *Nature* **345:**179–181.

Huber, R., Stoher, J., Hohenhaus, S., Rachel, R., Burggraf, S., Jannasch, H. W., and Stetter, K. O. 1995. *Thermococcus chitinophagus* sp. nov., a novel chitin-degrading, hyperthermophilic archeum from a deep-sea hydrothermal vent environment. *Arch. Microbiol.* **164:**255–264.

Huber, R., Wilharm, T., Huber, D., Trincone, A., Burggraf, S., Koenig, H., Rachel, R., Rockinger, I., Fricke, H., and Stetter, K. O. 1992. Aquifex pyrophilus gen. Nov. sp. nov., represents a novel group or marine hyperthermophilic hydrogen-oxidizing bacteria. *Syst. Appl. Microbiol.* **15:**340–351.

Jannasch, H. W., Wirsen, C. O., Molyneaux, S. J., and Langworthy, T. A. 1992. Comparative physiological studies on hyperthermophilic Archaea isolated from deep-sea hot vents with emphasis on *Pyrococcus* strain GB-D. *Appl. Environ. Microbiol.* **58:**3472–3481.

Jannasch, H. W., Wirsen, C. O., Molyneaux, S. J., and Langworthy, T. A. 1988. Extremely thermophilic fermentative archaebacteria of the genus *Desulfurococcus* from deep-sea hydrothermal vents. *Appl. Environ. Microbiol.* **54:**1203–1209.

Jeanthon, C., L'Haridon, S., Reysenbach, A.-L., Corre, E., Vernet, M., Messner, P., Sleytr, U. B., and Prieur, D. 1999. *Methanococcus vulanius* sp. nov., hyperthermophilic methanogen isolated from East Pacific Rise, and identification of *Methanococcus* sp. DSM 4213 as *Methanococcus fervens* sp. nov. *Int. J. Sys. Bacteriol.* **49:**583–589.

Jones, W. J., Leigh, J. A., Mayer, F., Woese, C. R., and Wolfe, R. S. 1983. *Methanococcus jannaschii* sp. nov., an extremely thermophilic methanogen from a submarine hydrothermal vent. *Arch. Microbiol.* **136:**254–261.

Jones, W. J., Stugard, C. E., and Jannasch, H. W. 1989. Comparison of thermophilic methanogens from submarine hydrothermal vents. *Arch. Microbiol.* **151:**314–318.

Kobayashi, T., Kwak, Y. S., Akiba, T., Kudo, T., and Horikoshi, K. 1994. *Thermococcus profundus* sp. nov., a new hyperthermophilic archaeon isolated from a deep-sea hydrothermal vent. *Syst. Appl. Microbiol.* **17**:232–236.

Kurr, M., Huber, R., Kônig, H., Jannasch, H. W., Fricke, H., Trincone., A., Kristjansson, J. K., and Stetter, K. O. 1991. *Methanopyrus kandleri*, gen. and sp. nov. represents a novel group of hyperthermophilic methanogens, growing at 110°C. *Arch. Microbiol.* **156**:239–247.

Kwak, Y. S., Kobayashi, T., Akiba, T., Horikoshi, K., and Kim, Y. B. 1995. A hyperthermophilic sulfur-reducing archaebacterium, *Thermococcus* sp. DT1331, isolated from a deep-sea hydrothermal vent. *Biosci. Biotech. Biochem.* **59**:1666–1669.

L'Haridon, S., Reysenbach, A.-L., Glenat, P., Prieur, D., and Jeanthon, C. 1995. Hot subterranean biosphere in a continental oil reservoir. *Nature* **337**:223–224.

L'Haridon, S., Cilia, V., Messner, P., Raguenes, G., Gambacorta, A., Sleytr, U. W., Prieur, D., and C. Jeanthon. 1998. *Desulfurobacterium thermolithotrophicum* gen. nov., sp. nov., a novel autotrophic, sulfur-reducing bacterium isolated from a deep-sea hydrothermal vent. *Int. J. Syst. Bacteriol.* **48**:701–711.

Legin, E., Ladrat, C., Godfroy, A., Barbier, G., and Duchiron, F. 1997. Thermostable amylolytic enzymes of thermophilic microorganisms from deep-sea hydrothermal vents. *C. R. Acad. Sci. Serie III* **320**:893–898.

Leuschner, C., and Antranikian, G. 1995. Heat-stable enzymes from extremely thermophilic and hyperthermophilic microorganisms. *World J. Microbiol. Biotechnol.* **11**:95–114.

Maher, K. A., and Stevenson, D. J. 1988. Impact frustration of the origin of life. *Nature* **331**:612–614.

Marteinsson, V. T., Birrien, J. L., Jeanthon, C., and Prieur, D. 1996. Numerical taxonomic study of thermophilic *Bacillus* isolated from three geographically separated deep-sea hydrothermal vents. *FEMS Microbiol. Ecol.* **21**:255–266.

Marteinsson, V. T., Birrien, J. L., Kristjansson, J. K., and Prieur, D. 1995. First isolation of thermophilic aerobic non-sporulating heterotrophic bacteria from deep-sea hydrothermal vents. *FEMS Microbiol. Ecol.* **18**: 163–174.

Marteinsson, V. T., Watrin, L., Prieur, D., Caprais, J.-C., Raguenes, G., and Erauso, G. 1995. Phenotypic characterization, DNA similarities, and protein profiles of twenty sulfur-metabolizing hyperthermophilic anaerobic Archaea isolated from hydrothermal vents in the Southwestern Pacific Ocean. *Int. J. Syst. Bacteriol.* **45**:623–632.

Marteinsson, V. T., Birrien, J. L., Reysenbach, A.-L., Vernet, M., Marie, D., and Prieur, D. 1996. Isolation of a hyperthermophilic archaeon under *in situ* pressure from a deep-sea hydrothermal vent and its pressure regulated protein. *Thermophiles* 96, September 1996, Athens, GA.

Marteinsson, V. T., Moulin, P., Birrien, J. L., Gambacorta, A., Vernet, M., and Prieur, D. 1997. Physiological responses to stress conditions and barophilic behavior of the hyperthermophilic vent archaeon *Pyrococcus abyssi*. *Appl. Environ. Microbiol.* **63**:1230–1236.

Moyer, C. L., Dobbs, F. C., and Karl, D. M. 1995. Phylogenetic diversity of the bacterial community from a microbial mat at an active, hydrothermal vent system, Loihi Seamount, Hawaii. *Appl. Environ. Microbiol.* **61**:1555–1562.

Moyer, C. L., Tiedje, J. M., Dobbs, F. C., and Karl, D. M. 1998. Diversity of deep-sea hydrothermal vent *Archaea* from Loihi Seamount, Hawaii. *Deep-Sea Res. II* **45**:303–317.

Neuner, A., Jannasch, H. W., Belkin, S., and Stetter, K. O. 1990. *Thermococcus litoralis* sp. nov.: a new species of extremely thermophilic marine archaeabacteria. *Arch. Microbiol.* **153**:205–207.

Pledger, R. J., and Baross, J. A. 1989. Characterization of an extremely thermophilic archaebacterium isolated from a black smoker polychaete (*Paralvinella* sp.) at the Juan de Fuca Ridge. *Syst. Appl. Microbiol.* **12**: 249–250.

Pledger, R. J., and Baross, J. A. 1991. Preliminary description and nutritional characterization of a chemoorganotrophic archaebacterium growing at temperatures of up to 110°C isolated from a submarine hydrothermal vent environment. *J. Gen. Microbiol.* **137**:203–211.

Pley, U., Schipka, J., Gambacorm, A., Jannasch, H. W., Fricke, H., Rachel, R., and Stetter, K. O. 1991. *Pyrodictium abyssi* sp. nov. represents a novel heterotrophic marine archaeal hyperthermophile growing at 110°C. *Syst. Appl. Microbiol.* **14**:243–245.

Prieur, D. 1992. Physiology and biotechnological potential of deep-sea bacteria. In Herbert, R. A., and Sharp. R. J. (eds.), *Molecular biology and biotechnology of extremophiles* (pp. 163–197). Glasgow and London: Blacki.

Prieur, D., Erauso, G., and Jeanthon, C. 1995. Hyperthermophilic life at deep-sea hydrothermal vents. *Planet. Space. Sci.* **43**:115–122.

Prieur, D., Erauso, G., Llanos, J., Deming, J. W., and Baross, J. A. 1992. Effect of hydrostatic pressure on mesophilic bacteria and ultra thermophilic archaebacteria from deep-sea hydrothermal vents. In *High pressure and biotechnology, Colloque Inserm 224*, pp. 19–25.

Raguenes, G., Meunier, J-R., Antoine, E., Godfroy, A., Caprais, J-C., Lesongeur, F., Guezennec, J., and Barbier, G. 1995. Biodiversité d'Archaea hyperthermophiles de sites hydrothermaux du Pacifique oriental. *C.R. Acad. Sci. Paris, Sciences de la vie/Life Sci.* **318:**395–402.

Reysenbach, A.-L., and Deming, J. W. 1991. Effects of hydrostatic pressure on the growth of hyperthermophilic archaebacteria from the Juan de Fuca Ridge. *Appl. Environ. Microbiol.* **57:**1271–1274.

Reysenbach, A. L., Longnecker, K., and Kirshtein, J. 2000. Novel bacterial and archaeal lineages from an in situ growth chamber deployed at a mid-Atlantic ridge hydrothermal vent. *Appl. Environ. Microbiol.* **66:** (in press).

Ritzau, M., Keller, M., Wessels, P., Stetter, K. O., and Zeeck, A. 1993. Now cyclic polysulfides from hyperthermophilic archae of the genus *Thermococcus*. *Liebigs Ann. Chem.* 871–876.

Shock, E. L. 1996. Hydrothermal systems as environments for the emergence of life. In *Evolution of hydrothermal ecosystems on Earth (and Mars?)*. Chichester, England: Wiley. (Ciba Foundation Symp. **202:**40–60.)

Stetter, K. O. 1996. Hyperthermophilic procaryotes. *FEMS Microbiol. Rev.* **18:**149–158.

Stetter, K. O. 1998. Hyperthermophiles: Isolation, classification and properties. In Horikoshi, K. and Grant, W. D. (eds.), *Extremophiles: Microbial life in extreme environments*. New York: Wiley-Liss.

Stetter, K. O., Huber, R., Blôch, E., Kurr, M., Eden, R. D., Fielder, M., Cash, H., and Vance, I. 1993. Hyperthermophilic archaea are thriving in deep North Sea and Alaskan oil reservoirs. *Nature* **365:**743–745.

Stevens, T. O. 1997. Subsurface microbiology and the evolution of the biosphere. In Amy, P., and Haldeman, D. (eds.) *The microbiology of the terrestrial subsurface* (pp. 203–221). Boca Raton, FL: CRC Press.

Von Damm, K. L. 1995. Controls on the chemistry and temporal variability of seafloor hydrothermal fluids. In Humphris, S. E., Zierenberg, R. A., Mullineaux, L. S. and Thomson, R. E. (eds.), *Seafloor hydrothermal systems: Physical, chemical, biological, and geological interactions* (pp. 222–247). Washington, D.C.: American Geophysical Union.

Wächtershäuser, G. 1998. Pyrite formation, the first energy source for life: A hypothesis. *Syst. Appl. Microbiol.* **10:**207–210.

Zhao, H., Wood, A. G., Widdel, F., and Bryant, M. P. 1988. An extremely thermophilic *Methanococcus* from a deep sea hydrothemal vent and its plasmid. *Arch. Microbiol.* **150:**178–183.

3

Biodiversity of Acidophilic Moderate Thermophiles Isolated from Two Sites in Yellowstone National Park and Their Roles in the Dissimilatory Oxido-Reduction of Iron

D. Barrie Johnson, Deborah A. Body, Toni A. M. Bridge, Debby F. Bruhn, and Francisco F. Roberto

1. INTRODUCTION

Bacteria that bring about dissimilatory transformations of iron are important from both biogeochemical and industrial perspectives (Ehrlich and Brierley, 1990; Johnson, 1995). The oxido-reduction of iron in extremely acidic ($pH < 3$) environments is particularly interesting because of the greater solubility of ionic (particularly ferric) iron and the relative stability of soluble ferrous iron under these conditions. Acidophilic iron-oxidizing bacteria are generally considered the most significant microorganisms in the biological processing of sulfide ores ("biomining") in which the accelerated oxidative dissolution of sulfidic minerals (e.g., pyrite, arsenopyrite, and chalcopyrite) solubilizes (e.g., copper) or releases (refractory gold) metals, thereby facilitating their recovery (Rawlings and Silver, 1995). Most research into bacterial iron transformations at low pH has focused on mesophilic chemolithotrophs, particularly *Thiobacillus ferrooxidans*, though a number of physiologically and phenotypically diverse mesophilic acidophiles, it is now known, are involved in

D. Barrie Johnson, Deborah A. Body, and Toni A. M. Bridge • School of Biological Sciences, University of Wales, Bangor, LL57 2UW, Wales. **Debby F. Bruhn and Francisco F. Roberto** • Lockheed Martin Idaho Technologies Co., Idaho National Engineering and Environmental Laboratory, Idaho, Falls, Idaho 83415-2203.

Thermophiles: Biodiversity, Ecology, and Evolution, edited by Reysenbach *et al.* Kluwer Academic / Plenum Publishers, New York, 2001.

the dissimilatory oxido-reduction of iron (Johnson, 1995; Norris and Johnson, 1997; Pronk and Johnson, 1992).

Certain acidophilic thermophiles can also oxidize ferrous iron. These may be divided into two groups (Norris, 1990): (1) "thermotolerant" (or "moderately thermophilic") isolates that tend to be gram-positive, rod-shaped bacteria whose temperature optima are about 50°C and that often display considerable metabolic diversity (e.g., *Sulfobacillus* spp. and *Acidimicrobium ferrooxidans*) and (2) "extreme thermophiles" whose temperature optima are 70–90°C, and which belong to the domain Archaea (Norris and Johnson, 1997). Although generally labelled as "iron-oxidizers," many of these thermophiles may also reduce ferric iron to ferrous under appropriate conditions. This trait is particularly common among thermotolerant iron bacteria (Ghauri and Johnson, 1991; Johnson et al., 1993) but has also been noted for the archaean *Sulfolobus acidocaldarius* (Brock and Gustafson, 1976).

Some of the thermal sites within Yellowstone National Park are extremely acidic (Brock, 1978) and therefore are potential sites for isolating novel strains of acidophilic thermophiles, including those that are involved in the biogeochemical cycling of iron. This paper reports the isolation and characterization of thermotolerant, acidophilic "iron bacteria" isolated from two sites in Yellowstone National Park and describes their physiological and phylogenetic biodiversity.

2. MATERIALS AND METHODS

2.1. Isolation of Acidophilic Microorganisms

Two sites were used as sources of iron-metabolizing bacteria and other acidophiles: (1) Frying Pan Hot Spring, a pool located near the roadside at the fringe of the Norris basin area; (2) the Sylvan Springs site which, although visible from the West Yellowstone-Old Faithful road, was less readily accessed and required crossing through the Gibbon River and a 1-kilometer hike through a wetland area. A single sample was taken at the Frying Pan Hot Spring in November 1989 by Dr. Daphne Stoner of the Idaho National Engineering and Environmental Laboratory; five samples were taken at different locations within the Sylvan Springs site during August 1993.

The methods for isolating microorganisms from both sites involved using solid media that had been developed specifically for culturing both mesophilic and moderately thermophilic acidophiles (Johnson, 1995), referred to as "direct isolation" in Table 1. Enrichment cultures of the sample from Frying Pan Hot Spring were set up in the following liquid media: pyrite (1%, w/v)/basal salts; 10 mM ferrous sulfate/0.025% tryptone soya broth; 10 mM glucose/basal salts. All media were poised initially at pH 2.0, and cultures were incubated, unshaken, at 45°C. After 7 days, culture samples were streaked onto ferrous iron overlay and yeast extract solid media. The water sample from Frying Pan Hot Spring was also inoculated directly onto ferrous iron overlay plates. Samples from Sylvan Springs were treated similarly, except that enrichment cultures were also prepared using a 1% (w/v) elemental sulfur/basal salts liquid medium (poised initially at pH 2.5) and ferrous iron/tetrathionate overlay plates were used to isolate acidophiles from both water samples and enrichment cultures.

Colonies of microorganisms were categorized initially on the basis of their mor-

Table 1
Characteristics of Two Sites Sampled within Yellowstone and the Diversity of Moderately Thermophilic Acidophilic Isolates Obtained from Them[a]

	Site					
	Frying Pan Hot Spring	Sylvan Springs (sampling point) 1	2	3	4	5
Temperature (°C)	50	61	40	80	68	40
pH	2.5	2.1	2.3	2.3	2.9	2.6
Iron-oxidizing isolates:						
(a) Via direct isolation	–	–	+ (YTF3)	–	+	+
(b) Via enrichment cultures	+ (YTF1)	+ (YTF17)	N.D.[b]	N.D.	N.D.	+ (YS5-F1) (YTF5)
Heterotrophic isolates:						
(a) Via direct isolation	–	–	+ (YTH16)	+ (YTH15)	+	+ (YTH14)
(b) Via enrichment cultures	+ (YTH1, 2, 3, & 5)	+	+ (YTH11 & 12)	+ (YTH13)	+	+ (YS5-H1)
Phototrophic isolates:						
Via direct isolation	–	–	+	–	–	+ (YTP2)

[a]Thermophilic sulfur-oxidizing acidophiles were also isolated from streaking samples from Sylvan Springs sites 1, 2, and 5 directly onto ferrous iron/tetrathionate plates.
[b]N.D.: not determined.

phological characteristics as iron oxidizers, sulfur oxidizers, heterotrophs, or phototrophs. Isolates were examined by phase contrast microscopy and subcultured in liquid media. These were 1% sulfur, pH 2.0 (sulfur oxidizers) and 10 mM ferrous sulfate/0.02% yeast extract, pH 2.0 (all other acidophiles). Isolates were purified by repeated plating and single colony isolation. Culture purity was checked periodically, again using specific solid media.

2.2. Measurement of Oxido-Reduction of Iron by Yellowstone Isolates

Cultures were grown routinely in media containing 25 mM ferrous sulfate/0.02% yeast extract/basal salts, occasionally supplemented with 10 mM glucose (or glycerol) and poised (initially) between pH 1.8 and 2.0 with 5 M H_2SO_4. Shake flask experiments were carried out in 250-mL Ehrlenmeyer flasks, varying the aeration status of cultures by agitation (or not) and by using different culture volumes. Isolates were also grown in a fermenter (LH Level 2 system, Inceltech, Reading, U.K.) in which temperature, dissolved oxygen, and pH were closely regulated. Changes in ferrous iron concentrations were monitored by titrating culture aliquots with 1 mM potassium permanganate in dilute sulfuric acid. Variations in the general culture conditions included replacing ferrous sulfate with ferric sulfate (at 25 mM), varying concentrations of yeast extract, and supplementing media with trace elements.

2.3. Determination of Specific Rates of Iron Oxidation and Reduction

The specific rates of iron oxidation and reduction (as amount of Fe^{2+} or Fe^{3+} transformed/unit time/unit quantity of cell protein) were measured for a single isolate (YTF1) from Frying Pan Hot Spring. Cultures were pregrown in 25 mM ferrous sulfate/10 mM glycerol/0.02% (w/v) yeast extract (pH 2.0) at dissolved oxygen (DO_2) concentrations of 80% or 20%, to the point at which all of the iron had been oxidized. To measure specific rates of iron oxidation, 90% of the culture volume was removed and replaced with fresh medium containing 25 mM ferrous sulfate, and changes in concentrations of ferrous iron (using the ferrozine assay (Lovley and Phillips, 1987)) and protein (using the Folin–Lowry method (Lowry et al., 1951)) were monitored for three hours. Specific rates of iron reduction were measured similarly, except that ferric sulfate was used in place of ferrous sulfate and the assay was carried out under anoxic conditions (by sparging cultures with nitrogen gas) for a 6-hour period.

2.4. Genomic DNA Isolation

Cultures of isolates YTH1, YTH2, YTF1, and YS5-F1 (500 mL) were grown in appropriate liquid media at pH 1.9, centrifuged at 500 rpm for 10 min in a Sorvall SS34 rotor in a Sorvall RC5B centrifuge (DuPont), and the resulting pellet washed once with 0.05 M H_2SO_4 and then with 10 mM Tris HCl, 1 mM EDTA, pH 8.0 (TE). the cell pellet was resuspended in 0.5 mL TE containing 0.6 mg/mL lysozyme supplemented with 50 mM EDTA, incubated for 1 h at 50°C, and then 0.1 mg/mL proteinase K, 1% SDS was added. After gentle inversion of the tube by hand, the mixture was incubated for an additional 1–3 h at 50°C without shaking, until the solution cleared. The DNA was extracted with phenol:chloroform:isoamyl alcohol (28:28:1, v/v/v) by gentle inversion for 10 min at room temperature. The biphasic mixture was separated by centrifugation at 10,000 rpm for 20 min at 4°C, and the upper aqueous layer containing DNA was carefully removed with a disposable polypropylene Pasteur pipet. Nucleic acids were precipitated upon addition of one-tenth volume 3M sodium acetate, pH 5.5, and 0.6 vol. isopropanol. The large molecular weight DNA precipitated as viscous, white strands that were spooled out with a sterile glass Pasteur pipet and pulled out to give a small diameter (ca 0.5 mM). The spooled DNA was washed with 70% ethanol and excess ethanol removed by briefly touching the tip of the pipet to the side of a sterile plastic microcentrifuge tube. The resulting genomic DNA was dissolved in 1 mL of TE at 4°C overnight without shaking.

2.5. PCR Amplification of 16S rRNA Genes

Nearly full-length 16S ribosomal DNA (rDNA) gene fragments were amplified from genomic DNA using dU-8(5′-ACGCGUACUAGUAGAGTTTGATCCTGGCTCAG-3′) and dU-1492R (5′-AUGGAGAUCUCUGGYTACCTTGTTACGACTT) primers, corresponding to *Escherichia coli* ribosomal RNA sequence positions 8–27 and 1510–1492, respectively. Modifications consisted of including the sequence "tails" containing deoxyuracil base substitutions to allow cloning in the selected vector system (see later). Amplification was performed using *Tfl* DNA polymerase (*Thermus flavus*; Epicentre Technologies, Madison, WI) and a Thermolyne PTC-100 thermocycler. In most cases, unique amplification products of about 1500 base pairs (bp) were obtained.

2.6. Cloning of Amplified Sequences

The amplified 16S rRNA gene sequences were cloned into the pAMP18 vector (BRL/ Life Technologies, Bethesda, MD) after uracil deglycosylase treatment according to manufacturer's instructions. Authentic clones were verified by restriction enzyme digestion to confirm that inserts were the expected size.

2.7. Sequencing of Cloned 16S rRNA Genes

Sequences of amplified and cloned 16S rDNA gene fragments were obtained by cycle sequencing of the double-stranded plasmid templates using thermocycle sequencing protocols and infrared dye end-labelled primers with a Li-Cor 4000L automated DNA sequencer (Li-Cor Inc., Lincoln, NE). In most cases, sequences represented the consensus of at least three separate sequence runs.

2.8. Sequence Analysis and Phylogenetic Tree Assembly

All data were compared with 16S rRNA sequences from the Ribosomal Database Project (RDP; Maidak et al., 1997). Sequences were first compared to the prokaryotic subset of small subunit rRNA sequences in the RDP using the SIMILARITY_RANK program. The most similar database sequences were obtained from the RDP in aligned format, and the new sequences were aligned manually with them using the GDE multiple sequence editor (Smith, 1994). Evolutionary distance matrix methods (least-squares additive distance tree method; LSADT; DeSoete, 1983) were used for initial phylogenetic determinations and were also compared with the results of parallel analyses using maximum likelihood algorithms (PHYLIP; Felsenstein, 1993). Analyses were performed on a Sun ULTRA 1 computer.

3. RESULTS

3.1. Moderately Thermophilic Acidophiles Isolated from the Yellowstone Sites

The pH and temperature characteristics of the single pool at Frying Pan Hot Spring and the five springs and streams at Sylvan Springs that were sampled are given in Table 1. Sites are also identified from which the various moderately thermophilic and acidophilic isolates were isolated, either directly from plating water samples or following enrichment in liquid media. Moderate thermophiles were isolated from all six sampling sites, including two at Sylvan Springs that had measured temperatures of 80 and 68°C. Sampling point 5 at Sylvan Springs was particularly interesting, both in terms of the numbers and of the range of acidophiles that were isolated from it; this was a stream that drained the thermal area which was colonized by filamentous, "streamer-like" microbial growths. Acidophilic, moderately thermophilic isolates obtained from this stream (and from the thermal spring at sampling point 2) included iron oxidizers, sulfur oxidizers, heterotrophic bacteria, and phototrophic isolates. The latter were particularly interesting because they grew as pale green colonies on iron overlay and iron/tetrathionate overlay plates that were incubated in the dark. These isolates were purified on solid media, as described before, and were identified as strains of *Cyanidium caldarium*, a rhodophyte that grows heterotrophically as well as phototrophically (Brock, 1978). Cultures of this unicellular alga were maintained in the same way as heterotrophic bacterial isolates, in 10 mM ferrous sulfate/0.02% yeast extract medium in an nonilluminated incubation; exposure of cultures to light resulted in a perceptible increase of green coloration, resulting from an increased synthesis of chlorophyll-*a*.

One culture, coded YS5, was obtained by subculturing a sample of Sylvan Springs stream water (from sampling point 5) in ferrous sulfate medium unamended with yeast extract. After several transfers, the composition of the culture was examined by plating onto selective solid media. Two bacteria were identified: one (coded YS5-F1) was a large (3–4 μm by 1 μm) spore-forming rod that produced small, ferric-iron-stained colonies on solid media; the other (coded YS5-H1) was a slender (3 μm by 0.5 μm) spore-forming rod that formed larger, unstained, cream-colored colonies on iron overlay plates. Isolate YS5-F1 was a *Sulfobacillus*-like iron oxidizer, but isolate YS5-H1 was obligately heterotrophic, though it was found subsequently that it can reduce ferric iron to ferrous. Numerical ratios between the two bacteria were around 1:1 in ferrous sulfate medium, but nearer 1 (YS5-F1):9 (YS5-H1) when the medium was supplemented with 0.02% yeast extract. The association between the two bacteria was highly stable; both were present following more than twenty subcultures through "inorganic" ferrous sulfate medium. Indeed, the ability of the iron-oxidizer YS5-F1 to grow in pure culture in either ferrous sulfate or pyrite media was considerably poorer than that of the mixed culture (data not shown).

Pure cultures of five iron-oxidizing moderate thermophiles were obtained, one (YTF1) from Frying Pan Hot Spring and four (YTF3, YTF5, YTF17 and YS5-F1) from Sylvan Springs (Table 2). These formed morphologically distinct colonies on overlaid solid media, though growth of all Sylvan Springs isolates on ferrous iron/tetrathionate medium was superior to that containing ferrous iron alone. All five isolates, it was found, also oxidize elemental sulfur; addition of yeast extract (0.02%) to sulfur cultures accelerated the rate of sulfur oxidation in most cases. Cells of isolate YTF3 tended to grow as long filaments during the exponential phase, were nonmotile, and showed no obvious endospore formation. In contrast, all other iron-oxidizing isolates tended to grow as single or paired rods, were generally motile, and formed terminal oval endospores. Four heterotrophic (non iron-oxidizing) isolates were obtained from Frying Pan Hot Spring, and eleven from Sylvan Springs (Tables 1 and 2); most were highly motile rods (2–3 μm or 4–5 μm long), and many isolates formed terminal endospores.

3.2. Oxidation of Ferrous Iron by Yellowstone Isolates

A trait of all of the iron-oxidizing isolates grown in ferrous sulfate/yeast extract liquid medium poised (initially) at pH 2.0 was that iron oxidation tended not go to completion. After an initial rapid phase of oxidation during the first 20 hours or so of the incubation, by which time 20% to 65% (depending on the isolate) of available ferrous iron had been oxidized, iron oxidation stopped and was arrested for the following 40–50 hours (Fig. 1).

Table 2
Some Morphological and Physiological Characteristics of Moderately Thermophilic, Acidophilic Microorganisms Isolated from the Two Yellowstone Sites

Isolate	Cell morphology	Spore formation	Iron oxidation	Iron reduction	Preliminary identification
YTF1	Rods	+	+	+	*Sulfobacillus* sp.
YTF3	Filamentous rods	(−)	+	(−)	*Acidimicrobium* sp.
YTF5	Rods	+	+	(−)	*Sulfobacillus* sp.
YTF17	Rods	+	+	(−)	*Sulfobacillus* sp.
YS5-F1	Rods	+	+	(−)	*Sulfobacillus* sp.
YTH1	Rods	+	−	+	*Alicyclobacillus* sp.
YTH2	Rods	+	−	+	*Alicyclobacillus* sp.
YTH3	Rods	+	−	+	*Alicyclobacillus* sp.
YTH5	Rods	+	−	+	*Alicyclobacillus* sp.
YTH11	Rods	+	−	+	*Alicyclobacillus* sp.
YTH12	Rods	−	−	+	—
YTH13	Rods	(+)	−	+	*Alicyclobacillus* sp.
YTH14	Rods	−	−	−	—
YTH15	Rods	−	−	−	—
YTH16	Rods	+	−	+	*Alicyclobacillus* sp.
YS5-H1	Rods	+	−	+	*Alicyclobacillus* sp.
YTP2	Cocci/oval (eukaryotic)	−	−	−	*Cyanidium caldarium*

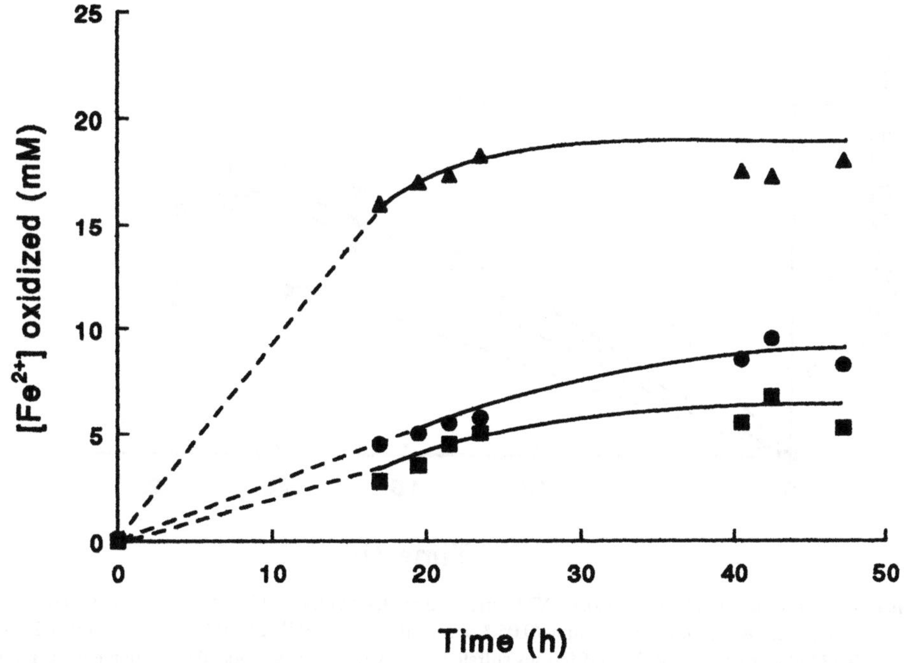

Figure 1. Oxidation of ferrous iron by moderately thermophilic isolates in media containing 25 mM ferrous sulfate and 0.02% (w/v) yeast extract. Symbols: YTF1 ▲; YTF17 ●; YS5-F1 ■.

Various experiments were conducted to account for this phenomenon, including examining whether the limited iron oxidation was due to some limiting organic or inorganic nutrient. It was noted, however, that starting cultures of strain YTF1 (and other isolates; data not shown) at a lower pH (1.8 rather than pH 2.0) resulted in complete iron oxidation (Fig. 2). This indicated that the reason for arrested iron oxidation by the moderate thermophiles was possibly inhibition by ferric iron, related to culture pH. To investigate this further, shake flask cultures of isolate YTF1 were grown in media containing 25 mM ferrous sulfate, 25 mM ferric sulfate and 0.02% yeast extract, poised either at pH 1.8 or 2.2, and changes in ferrous iron concentrations and pH were recorded. The results (Fig. 3) show that iron oxidation in the lower pH cultures began after a relatively short lag period and went rapidly to completion; in contrast, cultures poised initially at pH 2.2 displayed a longer lag period and did not begin to oxidize until the pH had dropped to 1.95 (these media are quite poorly buffered). The pH of these latter cultures increased slightly (to pH 2.05) during the following 40 hours and this again arrested iron oxidation, once more at about 60% of that available.

Culture doubling times (t_ds) of iron-oxidizing moderate thermophiles were evaluated from semilogarithmic plots of ferrous iron oxidized versus time. Isolates grown in 25 mM ferrous sulfate/0.02% yeast extract medium in shake flasks incubated at 50°C had the following t_ds: YTF1, 1.9 h; YTF3, 7.1 h; YTF5, 8.2 h; YTF17, 6.1 h; YS5-F1, 4.2 h. most of

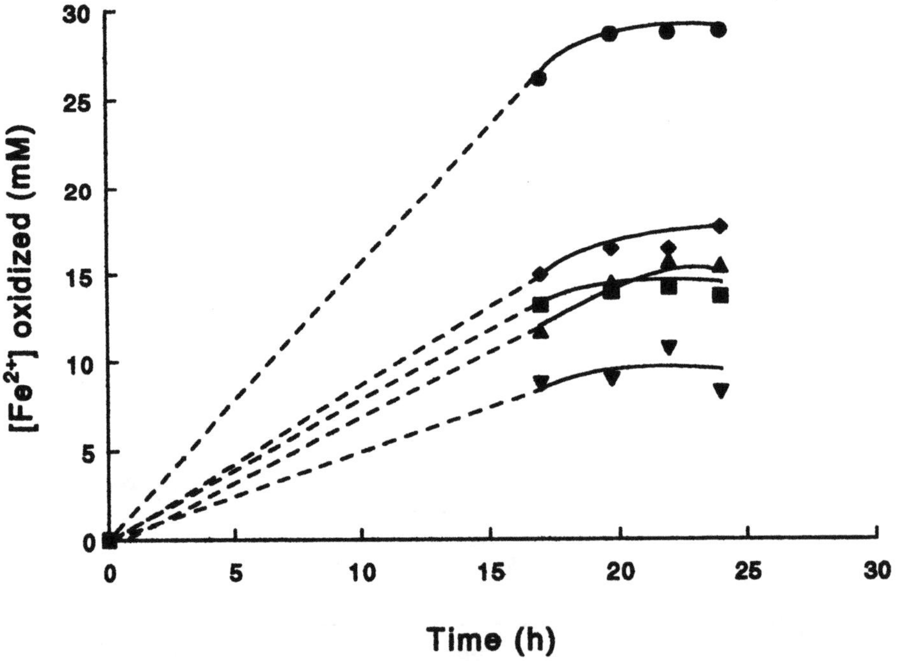

Figure 2. Ferrous iron oxidation by isolate YTF1 in liquid media. Symbols: 25 mM ferrous sulfate/0.02% (w/v) yeast extract, pH 2.0 ▲; 25 mM ferrous sulfate/0.05% yeast extract, pH 2.0 ▼; 25 mM ferrous sulfate/0.02% yeast extract/10 mM glucose, pH 2.0 ■; 25 mM ferrous sulfate/0.02% yeast extract/+ added trace elements, pH 2.0 ◆; 25 mM ferrous sulfate/0.02% yeast extract, pH 1.8 ●.

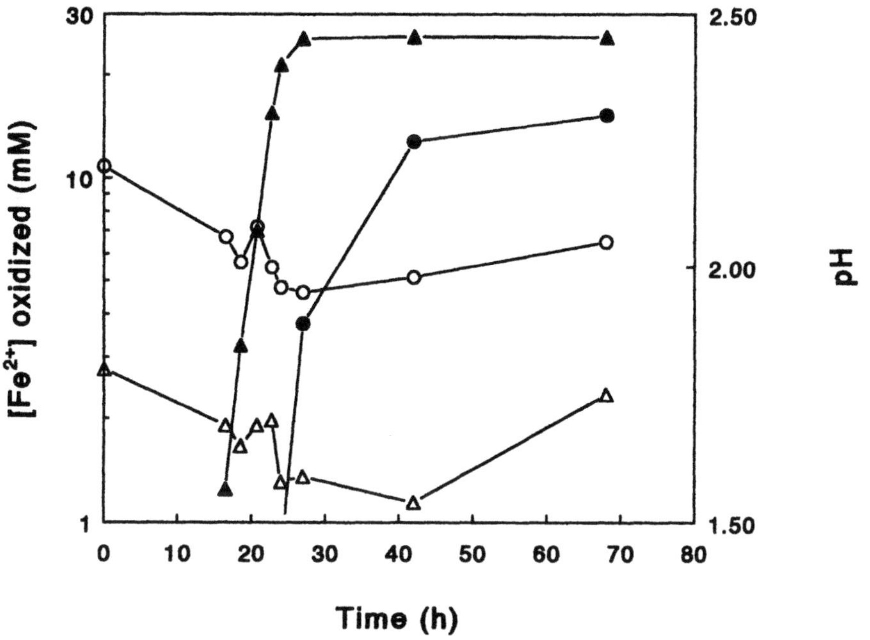

Figure 3. Relationship between ferrous iron oxidation (closed symbols) and culture pH (open symbols) by isolate YTF1 in media containing 25 mM ferrous sulfate, 25 mM ferric sulfate, and 0.02% (w/v) yeast extract. Symbols: medium poised initially at pH 2.2 ●, ○; medium poised initially at pH 1.80 ▲, △.

these data are comparable with published figures for other moderately thermophilic iron oxidizers (Ghauri and Johnson, 1991) except the isolate from Frying Pan Hot Spring, YTF1. When this isolate was grown under optimum conditions (fermenter culture; 25 mM ferrous sulfate/10 mM glycerol/0.02% yeast extract medium at 55°C and pH 2.0) growth rates were $0.51 \pm 0.07\ h^{-1}$ (equivalent to a mean culture doubling time of just over one hour). Growth of strain YTF1, measured either by increases in cell numbers (Fig. 4) or total protein (data not shown) was coupled to iron oxidation under these conditions. These data indicate that the moderate thermophile strain YTF1 is the fastest growing of all known iron-oxidizing microorganisms, including the thermophilic archaea, by a considerable margin (Norris and Johnson, 1997).

3.3. Reduction of Ferric Iron by Yellowstone Isolates

Isolate YTF1 reportedly reduces ferric iron when grown under microaerobic and anaerobic conditions (Ghauri and Johnson, 1991), a characteristic that was confirmed in the present study (data not shown). However, there was no evidence that any of the four Sylvan Springs isolates that oxidized ferrous iron could also reduce ferric iron. Specific rates of iron oxidation and iron reduction determined for isolate YTF1 pregrown under well aerated (80% DO_2) conditions showed that the propensity of this thermophile to oxidize ferrous iron was more than four times greater than that to reduce ferric iron (295 μg Fe^{2+} oxidized/minute/mg protein, compared to a mean of 62 μg Fe^{3+} reduced/minute/mg protein).

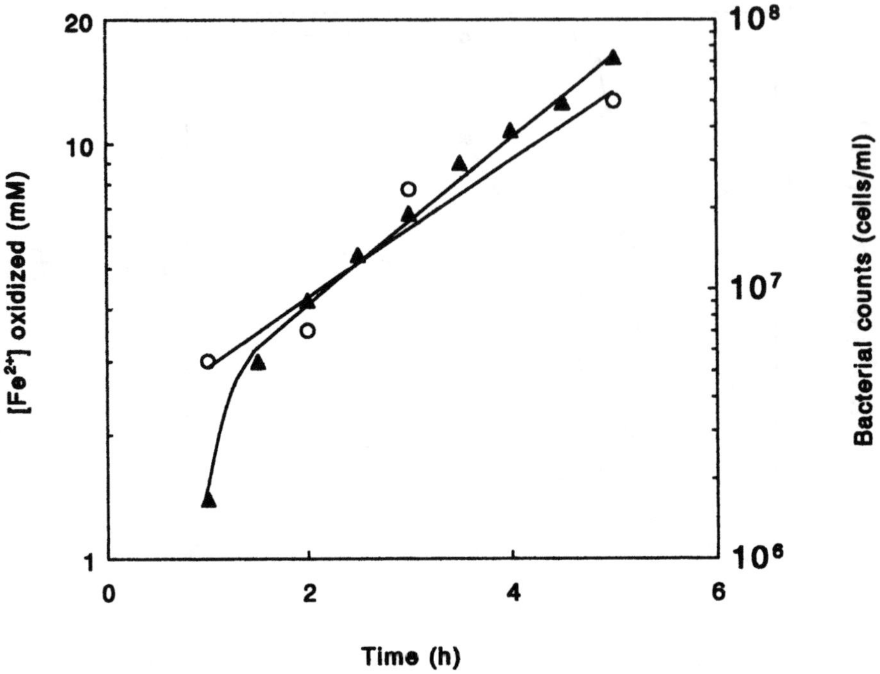

Figure 4. Correlation between ferrous iron oxidation (▲) and biomass increase (○), from plate counts of bacteria for isolate YTF1 grown in 25 mM ferrous sulfate/0.02% (w/v) yeast extract/10 mM glycerol medium.

Somewhat surprisingly, when YTF1 was grown at a lower oxygen concentration (20% DO_2), the mean specific rate of ferric iron reduction (18 μg Fe(III) reduced/minute/mg protein) was much less than that measured in cells grown at 80% DO_2. Specific iron oxidation was not estimated for cultures grown at 20% DO_2.

Most of the moderately thermophilic heterotrophic (i.e., non iron-oxidizing) isolates obtained from the two Yellowstone sites could reduce ferric iron to ferrous, though growth of these acidophiles was, in general, more readily inhibited by ferric iron (>10 mM) than the iron-oxidizing isolates. Enhanced tolerance of the non-iron-oxidizing moderate thermophiles to ferric iron was achieved by subculturing in media containing incrementally increasing concentrations of ferric sulfate. All of these isolates (except isolates YTH14 and 15) eventually grew in media (pH 2.0) containing 20 mM ferric sulfate. Growth rates of these isolates decreased with increasing concentrations of ferric iron (data not shown), and there was evidence that the toxicity of ferric iron to at least two isolates (strains YTH5 and YS5-H1) was greater at 50°C than at lower temperatures (Fig. 5).

3.4. Phylogenetic Analyses

Phylogenetic relationships between a subset of iron-oxidizing and heterotrophic isolates from the two Yellowstone sites and other moderate thermophiles (Lane et al., 1992), based on sequence analysis of 16S rDNA fragments, are shown in Fig. 6. For strains YTF1

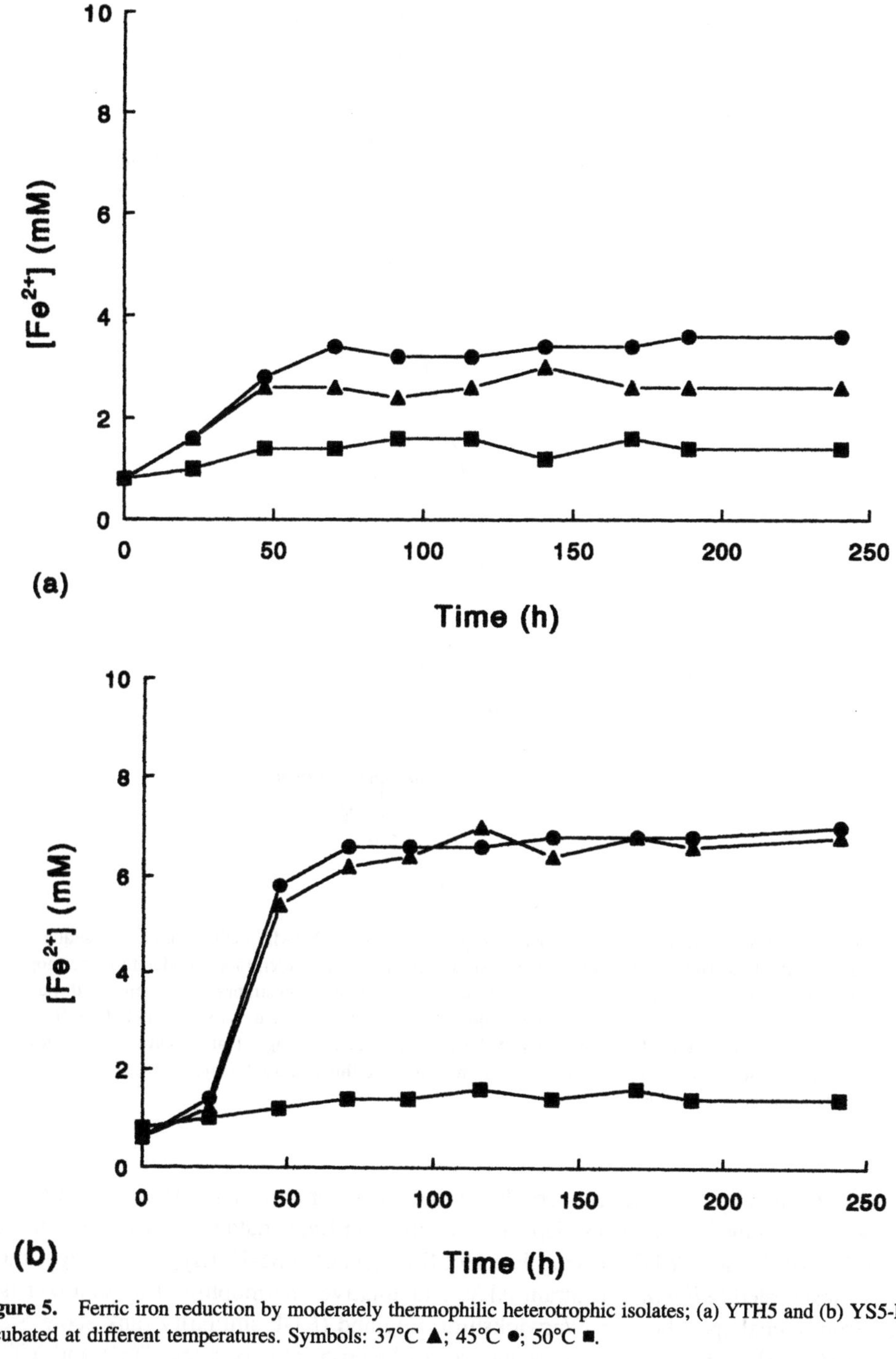

Figure 5. Ferric iron reduction by moderately thermophilic heterotrophic isolates; (a) YTH5 and (b) YS5-H1, incubated at different temperatures. Symbols: 37°C ▲; 45°C ●; 50°C ■.

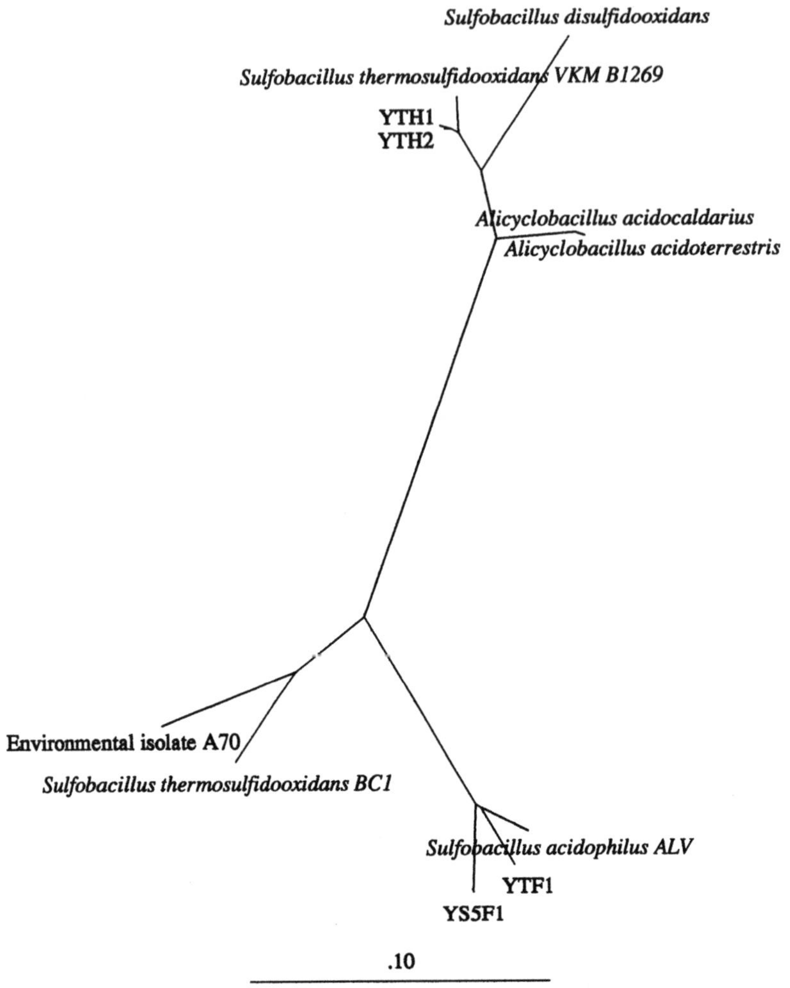

Figure 6. Rooted consensus tree (*A. acidoterrestris*) generated from DNAML analysis of aligned sequences of Yellowstone acidophilic isolates described in this study and previously identified, moderately thermophilic, acidophilic bacteria. Relative genetic divergence is indicated by the scale. Bootstrapped data were run through 100 replicate analyses, and the addition order of the organisms included in the analysis was jumbled ten times. All branches depicted were generated 100 times out of the 100 replicate samplings. Transition/reversion rates were varied between one and two to assess the impact on branching, and this tree used a rate of two (variation did not substantively alter the resulting tree).

and YS5-F1, 660 bp of sequence were determined, and for strains YTH1 and YTH2, 1286 bp were determined (sequences deposited in the GenBank database as accession nos. AF031643-6). Isolates YTF1 (Frying Pan Hot Spring) and YS5-F1 (Sylvan Springs) most closely resembled *Sulfobacillus* strain ALV, a facultative, thermophilic iron oxidizer isolated from a coal spoil heap in Warwickshire, England (RDP similarity rank score S_ab 0.681 and 0.551, respectively). The sequences of heterotrophic isolates YTH1 and YTH2 (both from Frying Pan Hot Spring) most closely resembled sequence data (Tuorova et al.,

1994) described for *Sulfobacillus thermosulfidooxidans* (S—ab 0.844 and 0.833, respectively). Phylogenetic relationships between these two Yellowstone isolates, the previously characterized moderately thermophilic and acidophilic heterotrophic bacteria *Alicyclobacillus acidocaldarius* and *Alicyclobacillus acidoterrestris* and the iron-oxidizer *S. thermosulfidooxidans* VKM B1269 are also shown in Fig. 6. It can be seen that this group forms a cluster which is separate from that containing iron-oxidizing strains YTF1, YS5-F1, ALV, and strain BC1 (an iron-oxidizing isolate from an English coal spoil that is considered very similar to *S. thermosulfidooxidans* VKM B1269). All fall within the *Alicyclobacillus* group, and their phylogenetic placement may coincide with their capacity to oxidize iron (there is currently some controversy regarding the sequence deposited for *S. thermosulfidooxidans* VKM B1269 as to whether it is representative the species or whether the BC1-derived sequence should be accepted as that of the type strain).

4. DISCUSSION

The Sylvan Springs area was described by Brock (1978) as "one of our favorite areas for acidophilic organisms"; data from springs recorded by Brock and his co-workers ranged from 45 to 91°C, from pH 1.8 to 8.3, and had iron concentrations of between <1 and 33 mg/L. This site also proved highly productive in terms of the numbers and diversity of moderately thermophilic acidophiles isolated in this study. Although we have focused primarily on bacteria involved in dissimilatory oxido-reduction of iron, the approach used also led to the concurrent isolation of other acidophiles, such as sulfur oxidizers and the eukaryotic phototroph *Cyanidium caldarium*. On several occasions, the latter organism has colonized acidic thermal areas within Yellowstone; what was surprising was its direct isolation on overlaid solid media (which was particularly successful on a medium that contained both ferrous iron and tetrathionate) on plates incubated in the dark. This was due to a combination of the metabolic versatility of the rhodophyte and the suitability of the overlay media for promoting growth of a variety of acidophilic microorganisms. The two sampling points (2 and 5) in the Sylvan Springs area from which *Cyanidium* was isolated were the coolest (at about 40°C) of those encountered. The other three sampling points (at 61–80°C) were seemingly too hot to support growth of this alga (Brock, 1978). Experiments with *Cyanidium* isolate YSP2 confirmed that it neither oxidized ferrous iron nor reduced ferric iron.

The iron-oxidizing moderate thermophiles isolated from the two sites in Yellowstone were originally differentiated from each other on the basis of colony morphologies on overlay medium, an approach that has also been used for mesophilic iron-oxidizing bacteria (Johnson, 1995). Subsequent analyses (e.g., of growth rates) confirmed that isolates were, at the minimum, different bacterial strains. In contrast to mesophilic iron-oxidizing bacteria such as *T. ferrooxidans*, all of these isolates showed limited (though variable) ferrous iron oxidation in cultures poised initially at pH 2.0. This was not a direct effect of pH because the isolates grew successfully at pH 2.0 both in media containing lower (5–10 mM) concentrations of iron or in sulfur media (data not shown). In the same way, the phenomenon was not due directly to inhibition by ferric iron (which accumulates as iron oxidation proceeds) because ferrous iron oxidation went to completion when cultures were poised initially at a lower pH. Ferric iron is known to be much more highly complexed than ferrous iron and forms stable complexes with ligands such as sulfate and hydroxide (Nordstrom et al., 1979). The dominant ferric iron complex in acidic solutions is $FeSO_4^+$, and other species such as $Fe(SO_4)_2^-$, $Fe(OH)^{2+}$, $Fe(OH)_2^+$, and free ferric iron (Fe^{3+}) assume minor proportions (Nordstrom et al., 1979). The ratios of these species in solution is, in part, mediated by pH, and it is possible that the observed effect was due to pH-related ferric iron inhibition caused by a specie(s) of ferric iron (such as $Fe(OH)_2^+$) which is more abundant at pH 2.0 than at pH 1.8.

Moderately thermophilic iron-oxidizing bacteria were first isolated (from an Icelandic thermal spring) and described in 1978, but relatively little work has been carried out on the systematics of this group of microorganisms. At present, there are two recognized genera of such bacteria. *Sulfobacillus* spp. (Norris et al., 1996) are gram-positive, spore-forming acidophiles, currently comprising two species (*S. thermosulfidooxidans* and *S. acidophilus*) that may be differentiated on some physiological (e.g., autotrophic growth on elemental sulfur) and molecular (e.g., mole % guanine + cytosine contents of chromosomal DNA) characteristics. Of the Yellowstone isolates described in this chapter, strains YTF1, YTF5, YTF17, and YS5-F1 all appear, from their morphological and physiological characteristics, to be *Sulfobacillus* spp.; sequence analysis of 16S rDNA indicates that strains YTF1 and YS5-F1 are more closely related to *S. acidophilus* strain ALV than to other sequenced thermophiles (Lane et al., 1992). At present, the other currently designated genus of iron-oxidizing moderate thermophiles, *Acidimicrobium*, is comprised of a single species, *A. ferrooxidans*, which branches quite separately within the gram-positive eubacterial division from *Sulfobacillus* spp. (Lane et al., 1992). The distinct cellular morphology of Sylvan Springs isolate YTF3 (fine rods that tend to form filaments, and no obvious evidence of endospore production) is, however, similar to that described for *A. ferrooxidans* strain TH3. Future work with isolate YTF3, which will include 16S rDNA sequence analysis, will establish whether this isolate is a novel strain of *A. ferrooxidans*.

The situation with the (non-iron-oxidizing) moderately thermophilic heterotrophic isolates is somewhat ambiguous. On the basis of their physiologies and morphologies, the Yellowstone isolates appear to belong to the genus *Alicyclobacillus* (Wisotzkey et al., 1992). However, 16S rDNA sequence similarities suggest that both Frying Pan Hot Spring isolates YTH1 and YTH2 are strains of *S. thermosulfidooxidans* (Tuorova et al., 1994). The proximity of this cluster to *Alicyclobacillus spp.*, and the distance from iron-oxidizing moderate thermophiles, including strain BC1 (which is considered to be a strain of *S. thermosulfidooxidans*) suggest that reclassifying the type strain of *S. thermosulfidooxidans* should be considered.

A novel feature noted in this work was the ability of moderately thermophilic heterotrophic acidophiles to reduce ferric iron. Previous data had indicated that isolates from Frying Pan Hot Spring could not reduce ferric iron when grown on solid media containing 25 mM ferric sulfate (Johnson and McGinness, 1991). This apparent discrepancy may have resulted either from the fact that lower concentrations of ferric iron were used in this study (growth of these bacteria were inhibited by concentrations of ferric iron of >10 mM) or because a different electron donor (glycerol) was used. Data from this study showed that most heterotrophic isolates (except the somewhat less acidophilic strains YTH14 and YTH15) grow at pH 2.0 and reduce soluble ferric iron to ferrous under conditions which were microaerobic rather than strictly anaerobic. The mechanism of ferric iron reduction by these bacteria has not been ascertained, though in mesophilic acidophiles iron reduction is mediated via anaerobic respiration rather than fermentation (Pronk and Johnson, 1992).

Moderately thermophilic iron-oxidizing bacteria, including Frying Pan Hot Spring isolate YTF1, reportedly catalyze the dissimilatory reduction of ferric iron under conditions of low DO_2 (Ghauri and Johnson, 1991). No unequivocal data, however, showed iron reduction by the four Sylvan Springs iron-oxidizing isolates. In addition, comparison of the specific rates of ferrous iron oxidation and ferric iron reduction by isolate YTF1 validate its primary description as an "iron oxidizer." The lower specific rate of ferric iron reduction recorded when this isolate was grown under 20% DO_2 compared with 80% DO_2 was intriguing; it is possible that the same enzyme system is involved in both iron oxidation and reduction in this acidophile, though specific ferrous iron oxidation rates in cultures grown under 20% DO_2 were not estimated. The ability to reduce as well as to oxidize iron has also been noted for the most well known acidophile, *T. ferrooxidans* (Pronk and Johnson, 1992).

The use of thermophilic (as opposed to mesophilic) metal-mobilizing bacteria in biomining operations, it has been suggested, is potentially advantageous because mineral ore leaching proceeds more rapidly at elevated temperatures (Ehrlich and Brierley, 1990). Sulfide mineral oxidation is exothermic, and temperatures within ore piles (heap leaching) and in large-scale tank reactors may frequently exceed 40°C. Rates of growth and iron oxidation by isolate YTF1 are more rapid than those recorded for all other metal-mobilizing acidophiles, and therefore this isolate is a potential candidate for ore bioprocessing. However, the growth of YTF1 in organic-free cultures is much slower than in yeast extract-containing media, and a requirement for extraneous organic substrates would seriously diminish the attraction of using microorganisms such as isolate YTF1 for mineral processing. However, moderately thermophilic acidophiles may form highly stable mixed cultures (such as culture YS5 from Sylvan Springs) that may promote more efficient iron oxidation than pure cultures (Norris et al., 1996). Therefore, consortia of moderate thermophiles, rather than pure cultures, might be more appropriate for industrial-scale mineral bioprocessing (Clark and Norris, 1996). The ability of isolates such as the iron-oxidizer YTF1 and the thermophilic heterotrophs to catalyze the dissimilatory reduction of ferric iron to ferrous may also have potential in removing secondary oxidation products, such as jarosites, that tend to form passivation layers on mineral surfaces during mineral processing. Recent work has shown that isolate YTF1 can accelerate the reductive dissolution of a range of ferric iron-containing minerals (Bridge and Johnson, 1998).

5. SUMMARY

A variety of acidophilic moderate thermophiles were isolated from two sites in Yellowstone National Park by direct isolation on selective solid media and from enrichment cultures. Isolates included iron- and sulfur-oxidizing bacteria, obligately heterotrophic bacteria, and *Cyanidium caldarium*-like rhodophytes. Phylogenetic analysis (16S rRNA gene sequencing) of two of the five strains of iron-oxidizing bacteria isolated showed that they were most closely related to *Sulfobacillus acidophilus*. The growth rate of one isolate (YTF1), grown under optimum conditions, was more rapid than that recorded for all other iron-oxidizing prokaryotes; the minimum mean generation time for this isolate was just over 1 hour. All of the iron-oxidizing isolates displayed incomplete oxidation of 25 mM ferrous iron when grown in ferrous sulfate/glycerol/yeast extract medium poised at pH 2.0, though when the initial pH of this medium was adjusted to pH 1.8, iron oxidation went to completion. It was postulated that this phenomenon might be due to inhibition of the bacteria by a ferric iron complex (such as $Fe(OH)_2^+$), produced as a result of iron oxidation, which was more abundant at the higher pH. Most of the obligately heterotrophic isolates could reduce ferric iron to ferrous. The capacity of this group of isolates to tolerate ferric iron and to reduce ferric iron was temperature-related. Most of the heterotrophic isolates which did not oxidize ferrous iron were from their physiological characteristics, *Alicyclobacillus*-like isolates, though phylogenetic analysis of two isolates places those organisms closest to the iron-oxidizing moderate thermophile *Sulfobacillus thermosulfidooxidans* VKM B1269 and the disulfide-, pyrite-, and sulfur-oxidizing bacterium, *Sulfobacillus disulfidooxidans*. Taken together, the physiological and phylogenetic characteristics of the bacteria isolated in this study are consistent with features of other members of the *Alicyclobacillus* group, of which a number were, coincidentally, first isolated from Nymph Creek in Yellowstone National Park.

ACKNOWLEDGMENTS. Toni Bridge is grateful to the Biotechnology and Biological Sciences Research Council (U.K.) for providing a research studentship. Work at the INEEL was supported by the U.S. Department of Energy through DOE Idaho Office Contract DE-AC07-94ID013223 to Lockheed Martin Idaho Technologies Company.

REFERENCES

Bridge, T. A. A., and Johnson, D. B. 1998. Reduction of soluble iron and reductive dissolution of ferric iron-containing minerals by moderately thermophilic iron-oxidizing bacteria. *Appl. Environ. Microbiol.* **64**:2181.

Brock, T. D. 1978. *Thermophilic microorganisms and life at high temperatures*. New York: Springer-Verlag.

Brock, T. D., and Gustafson, J. 1976. Ferric iron reduction by sulfur- and iron-oxidizing bacteria. *Appl. Environ. Microbiol.* **32**:567.

Clark, D. A., and Norris, P. R. 1996. *Acidimicrobium ferrooxidans* gen. nov., sp. nov.: Mixed culture ferrous iron oxidation with *Sulfobacillus* species. *Microbiology* **141**:785.

De Soete, G. 1983. A least squares algorithm for fitting additive trees to proximity data. *Psychometrika* **48**:621.

Ehrlich, H. L., and Brierley, C. L. (eds.). 1990. *Microbial mineral recovery*. New York: McGraw-Hill.

Felsenstein, J. 1993. PHYLIP (Phylogeny Inference Package), version 3.5c. Seattle: Department of Genetics, University of Washington.

Ghauri, M. A., and Johnson, D. B. 1991. Physiological diversity amongst some moderately thermophilic iron-oxidising bacteria. *FEMS Microbiol. Ecol.* **85**:327.

Johnson, D. B. 1995. Mineral cycling by microorganisms: Iron bacteria. In Allsop, D., Hawksworth, D. L., and Colwell, R. R. (eds.), *Microbial diversity and ecosystem function* (pp. 137–159). Wallingford, U.K.: CAB International.

Johnson, D. B. 1995. Selective solid media for isolating and enumerating acidophilic bacteria. *J. Microbiol. Meth.* **23**:205.

Johnson, D. B., and McGinness, S. 1991. Ferric iron reduction by acidophilic heterotrophic bacteria. *Appl. Environ. Microbiol.* **57**:207.

Johnson, D. B., Ghauri, M. A., and McGinness, S. 1993. Biogeochemical cycling of iron and sulphur in leaching environments. *FEMS Microbiol. Rev.* **11**:63.

Lane, D. J., Harrison, A. P., Jr., Stahl, D., Pace, B., Giovannani, S. J., Olsen, G. J., and Pace, N. R. 1992. Evolutionary relationships among sulfur- and iron-oxidizing eubacteria. *J. Bacteriol.* **174**:269.

Lovley, D. P., and Phillips, E. J. P. 1987. Rapid assay for microbially reducible ferric iron in aquatic sediments. *Appl. Environ. Microbiol.* **53**:1536.

Lowry, O. H., Rosenbrough, N. J., Farr, A. L., and Randall, R. J. 1952. Protein measurement with the Folin phenol reagent. *J. Biol. Chem.* **193:**265.

Maidak, B. L., Olsen, G. J., Larsen, N., Overbeek, R., McCaughey, M. J., and Woese, C. R. 1997. The RDP (Ribosomal Database Project). *Nucl. Acids Res.* **25:**109.

Nordstrom, D. K., Everett, E. A., and Ball, J. W. 1979. Redox equilibria of iron in acid mine waters. In Jenne, E. A. (ed.), *Chemical modeling in aqueous systems* (pp. 51–79). Washington, D.C.: American Chemical Society.

Norris, P. R. 1990. Acidophilic bacteria and their activity in mineral sulphide oxidation. In Ehrlich, H. L., and Brierley, C. R. (eds.), *Microbial mineral recovery*, New York: McGraw-Hill.

Norris, P. R., and Johnson, D. B. 1997. Acidophilic microorganisms. In K. Horikoshi, K., and Grant, W. D. (eds.), *Extremophiles: Microbial life in extreme environments* (pp. 133–154). New York: Wiley.

Norris, P. R., Clark, D. A., Owen, J. P., and Waterhouse, S. 1996. Characteristics of *Sulfobacillus acidophilus* sp. nov. and other moderately thermophilic mineral sulphide-oxidizing bacteria. *Microbiology* **141:**775.

Pronk, J. T., and Johnson, D. B. 1992. Oxidation and reduction of iron by acidophilic bacteria. *Geomicrobiol. J.* **10:**153.

Rawlings, D. W., and Silver, S. 1995. Mining with microbes. *Bio/Technology* **13:**773.

Smith, S. 1994. Genetic Data Environment, version 2.2. Champaign-Urbana: University of Illinois.

Tuorova, T. P., Poltoraus, A. B., Lebedeva, I. A., Tsaplina, I. A., Bogdanova, T. I., and Karavaiko, G. I. 1994. 16S ribosomal RNA (rDNA) sequence analysis and phylogenetic position of *Sulfobacillus thermosulfidooxidans*. *Syst. Appl. Microbiol.* **17:**509.

Wisotzkey, J. D., Jurtshuk, P., Jr., Fox, G. E., Deinhard, G., and Poralla, K. 1992. Comparative sequence analysis on the 16S RNA (rDNA) of *Bacillus acidocaldarius*, *Bacillus acidoterrestris* and *Bacillus cycloheptanicus* and proposal for creation of a new genus, *Alicyclobacillus* gen. nov. *Appl. J. Syst. Bacteriol.* **42:**263.

4

Presence of Thermophilic *Naegleria* Isolates in the Yellowstone and Grand Teton National Parks

Robert F. Ramaley, Pamela L. Scanlan, and William D. O'Dell

1. INTRODUCTION

During a study of the extremely thermophilic bacteria of Yellowstone, we became aware of the published report (Scaglia et al., 1983) of the isolation of a thermophilic amoeba (*Naegleria australiensis*) from the bathing pools of a commercial Italian hot spring spa and their observation that some of these newly isolated amoebae strains were as virulent for mice as *Naegleria fowleri*, the etiological agent of primary amoebic meningoencephalitis in humans (John, 1982; Marciano-Cabral, 1988). Because of the possibility that similar types of thermophilic pathogenic amoebae might be present in the Yellowstone and Grand Teton National Park thermal areas and possible exposure to park visitors and employees, we conducted a survey for *Naegleria* isolates, concentrated on those pools or outflows where it was known that visitors and employees used these pools for unauthorized bathing and soaking (Whittley, 1995).

Unlike contact with other water born pathogenic protozoans (e.g., *Giardia lamblia*, *Entamoeba histolitica*, and *Cryptosporidum*) (Mitchell et al., 1988), simple contact or drinking water containing the *Naegleria* is not sufficient to cause disease. A fatal exposure

Robert F. Ramaley, Pamela L. Scanlan, and William D. O'Dell • Department of Biochemistry and Molecular Biology, University of Nebraska Medical Center and Department of Biology, University of Nebraska at Omaha, Omaha, Nebraska 68198-4525.

Thermophiles: Biodiversity, Ecology, and Evolution, edited by Reysenbach *et al.* Kluwer Academic / Plenum Publishers, New York, 2001.

requires that a sufficient quantity of water/sediment be inhaled through the nose. Then, the amoebae penetrate the nasal mucosa and cribiform plate and travel along the olfactory nerves to the brain. The disease is rapidly fatal and usually causes death within 72 hours after the onset of symptoms, such as severe frontal headache (John, 1982; Marciano-Cabral, 1988; Martinez and Visvesvara, 1997).

The presence of *Naegleria* in either Yellowstone or Grand Teton National Parks had not previously been documented, although Boyle et al. (1982) obtained a thermophilic *Naegleria* isolate from a hot spring in Cody, Wyoming. Their isolate was not pathogenic in mice.

2. MATERIALS AND METHODS

2.1. Collection and Isolation of *Naegleria* Isolates

Single samples of approximately 750 mL were taken at each site, usually along with a small portion of algal mat. All samples were treated within 3 hours of collection. The samples were shaken vigorously before 100 to 150 mL was filtered through a 5.0-μm membrane filter (Millipore, Type SM) in a sterile disposable filter unit (Nalgene). The filters were placed upside down on the surface of 1.5% nonnutrient agar (NNA) plates that had been spread with a distilled water suspension of *Escherichia coli* B/r or *Thermus aquaticus* YT-1.

The plates were incubated at 45°C and observed daily for the appearance of amoebic plaques at the edge of the filters. Plaques were transferred by agar blocks to fresh NNA with bacteria for maintenance. The ability to form flagellates was tested by transferring agar blocks of actively growing amoebae to tubes of sterile Page's Amoeba Saline (PAS) (O'Dell and Stevens, 1973). All flagellate tests and incubations subsequent to isolation were done at 37°C. Strains were isolated by diluting a suspension of cysts in sterile distilled water so that samples spread on the surface of NNA plates produced between five and ten plaques (Jensen and Dubes, 1962). Clumping of cysts was monitored by phase-contrast microscopy so that each plaque was assumed to be a clone. Identification of the *Naegleria* isolates was made on the basis of accepted morphological criteria, the formation of flagellates at 37°C, and the observation of promitosis in Gomori trichrome stained slides (Page, 1967).

Axenic growth in several media was attempted for all of the *Naegleria* isolates. These media included the casitone-serum medium (CAS) of Cerva (Cerva, 1969), American Type Culture Collection (ATTC) medium 1034, the casitone-glucose-vitamins (CGVS) medium of Willaert (1971), and the serum-casein-glucose-yeast extract medium (SCGYEM) of DeJonckheere (1977). All of the media were modified by adding 200 units of penicillin and 200 μg of streptomycin/mL. Agar blocks were transferred from actively growing cultures on NNA plates and added directly to the axenic medium. Additional attempts, if necessary, were made by transferring cells from established axenic cultures to unsuccessful media.

2.2. Determination of Virulence of the *Naegleria* Isolates

Virulence in mice was determined by intranasal (IN) and intracerebral (IC) inoculation of axenic amoebae into CF-1 (Sasco) weaning (12–15 grams) female mice. The mice were anesthetized with 25 μl ketamine (Ketaset) given intramuscularly before the inoculation of 10 μl IN or 20 μl IC with a 26 g needle. *Naegleria fowleri* (ATCC 30863) was used for infected controls, and SCGYEM medium was given to the uninfected controls. The mice were observed daily for 21 days for the appearance of symptoms. The brains of the dead mice and of the survivors were removed, and a portion was transferred to NNA spread with *E. coli* B/r which was then incubated at 37°C and observed for amoebic growth.

2.3. Determination of Growth of *Naegleria* on *Thermus* Strains

The ability of selected strains to use various thermophilic bacteria as a source of food was determined by adding axenic amoebae to the surface of nonnutrient agar plates that had been spread with the *Thermus* strains and by incubating the cultures at 35 or 44°C. The *Thermus* strains were grown in 300-mL sidearm flasks (50 mL of 0.3% yeast extract, 0.3% tryptone in Castenholtz salts (pH 8.0) (Ramaley and Hixson, 1970)). Forty mL of the culture was centrifuged (5,000 × g for 10 minutes), the pellet was resuspended in sterile distilled water, and the cells were centrifuged again. The washed cell pellet was resuspended in sterile distilled water at 0.1 g (wet weight) per mL, and 0.1 mL of the cell suspension was spread on plates on nonnutrient agar containing Page Amoeba Saline (O'Dell and Stevens, 1973), and nonnutrient agar containing Castenholtz salts (Ramaley and Hixon, 1970), adjusted to pH 8.0. *Naegleria lovaniensis* (30569/TS-1), *N. australiensis* (30958/87), and *N. fowleri* (30863/83) cultures (2.4 mL) were centrifuged in two 1.2-mL microcentrifuge tubes in a Beckman microcentrifuge, and each pellet was resuspended in 0.75 mL. A 4-μL quantity of suspension was plated on each of these plates (three applications per plate) and incubated at the indicated temperatures.

3. RESULTS

3.1. Isolation of Thermophilic Amoebae from the Yellowstone/Grand Teton Ecosystem

Table 1 lists the sample locations where thermophilic amoebae isolates were obtained. The highest number of amoebae was found in the Huckleberry Hot Springs and the nearby springs along Polecat Creek in the north part of Grand Teton National Park (near Flagg Ranch). These springs have a much lower outflow temperatures 40–60°C than most of the Yellowstone hot springs and are extensively used by cross-country skiers for bathing.

Eleven of the fifteen initial sample sites gave thermophilic amoebae isolates (Table 1). All of the positive sites were in neutral to slightly alkaline springs. No thermophilic amoebae were isolated from acidic springs such as Frying Pan Springs or from hot springs or runoff channels at temperatures above 47°C. Of the initial thirty-four isolates, sixteen were identified as *Naegleria* based on morphological appearance, the formation of flagellates at 37°C, and observation of promitosis in Gomori trichrome stained slides (Page, 1967) and were present in eight of the eleven thermophilic amoebae-positive sample sites. Two strains of *Hartmannella vermiformis* were isolated, one from Twin Falls and one from a commercial hot spring and spa located outside of the park. The remaining sixteen isolates could not be further identified, but they were not typical soil amoebae.

Table 1
Thermophilic Amoeba Isolates from Hot Springs

Sample site	Water temp. (°C)	*Naegleria*/ total isolates	Media			
			SCGYEM	CAS	CGVS	1034
Commercial hot spring A	33	2/3	2/2	2/2	0	2/2
Commerical hot spring B	38	0/1	—	—	—	
Commerical hot spring pool B	34	2/4	0/2	0/2	0/2	0/2
Boiling River	46	1/3	0/1	1/1	1/1	1/1
Frying Pan Runoff	33	0	—	—	—	
Artist's Paint Pot	35	0	—	—	—	
Firehole River	22	2/3	0/2	1/2	1/2	1/2
Solitary Spring	46	1/1	0/1	0/1	0/1	0/1
Twin Butte Spring	42	0/3	—	—	—	
Small Spring (White Creek)	46	0/1	—	—	—	
Pine Spring (White Creek)	62	0	—	—	—	
Twin Falls (White Creek)	47	1/3	0/1	1/1	1/1	1/1
Ivory Geyser Runoff	32	0	—	—	—	
Ranger pool (Old Faithful)	40	4/9	2/4	2/4	2/4	3/4
Huckleberry Springs	45	3/3	2/3	3/3	3/3	3/3
Totals		16/34	6/16	10/16	8/14	11/16

The results of the axenic cultivation attempts are also listed in Table 1. Six of the *Naegleria* isolates grew in SCGYEM media which is a presumptive growth media for *N. fowleri* (ATCC 30863) type isolates. Medium 1034 supported the axenic growth of more *Naegleria* isolates than any of the other media. Four of the *Naegleria* isolates failed to grow in any of the axenic media.

Intercerebral inoculation of strain BHS 3-1 amoebae isolated from a commercial hot spring bathing and swimming pool area located 45 miles north of the Yellowstone National Park boundary resulted in the death of three of ten mice (Table 2). Eight of ten mice infected intracerebrally with strain HS-2 amoebae (isolated from Huckelberry Hot Spring) had symptoms characteristic of the disease on days 3 to 7 postinfection, but recovered. None of the isolates tested were infective by the intranasal route. Amoebae were recovered from the brains of all of the dead mice but not from the survivors. Thus, none of the thermophilic isolates were as pathogenic as the type species of *N. fowleri*. The isolates from the commercial hot spring and Huckleberry Hot Springs are *N. australiensis* as determined by the pattern of isoelectric focusing of the cellular enzymes and the use of substrate test kits (API ZYM systems) (Scanlan, 1988).

One of the questions aroused by the isolation of these thermophilic amoebae was what serves as the natural food source of these organisms. Figure 1, a scanning electron micrograph of the pathogenic *N. fowleri* (ATCC 30863), shows the sucker-like structures (ameobastome) (John et al., 1984) by which *Naegleria* engulfs bacteria. The conventional laboratory food source for *Naegleria* is a lawn of *Escherichia coli* cells on nonnutrient agar. The results shown in Table 1 were obtained by using an *E. coli* lawn. Because it is likely that *E. coli* is not a natural component of the microbial flora of hot springs and outfall channels, we investigated the use of various *Thermus* cells to serve in place of *E. coli* as a food source

Table 2
Virulence of *Naegleria* Isolates

Strain	Route[a]	Dose	Dead/total	Day of death[b]
BHS 3-1	IC	86,000	3/10	1, 3, 5, S, S, S, S, S, S, S
BHS 3-1	IN	43,000	0/10	No deaths
BHS 4-1	IC	42,000	0/10	No deaths
BHS 4-1	IN	21,000	0/10	No deaths
HS-1	IC	100,000	0/10	No deaths
HS-1	IN	58,000	0/10	No deaths
HS-2	IC	200,000	0/10	No deaths
HS-2	IN	100,000	0/10	No deaths
N. fowleri	IC	92,000	10.10	4, 4, 4, 4, 4, 4, 5, 5, 5, 5
N. fowleri	IN	46,000	10/10	7, 8, 8, 8, 8, 9, 9, 9, 9, 9

[a]Intercerebral (IC) or internasal (IN) administration of *Naegleria*.
[b]Number indicates day of death post infection for each mouse; S = survived.

for *Naegleria*. Table 3 shows the growth of *N. fowleri*, *N. australiensis*, and *N. louvaniensis* on *T. aquaticus* YT-1 (Brock and Freese, 1969), *Thermus* X-1 [a non–pigmented strain of "*Thermus*" (Ramaley and Hixson, 1970)], *Thermus* S-1 (a non–pigmented strain isolated from a hot water tank at Dana College in Blair, Nebraska by Stan Johnson), K-2 isolate (a red–pigmented strain similar to *Thermus ruber* (Loginova and Egorova, 1975) isolated

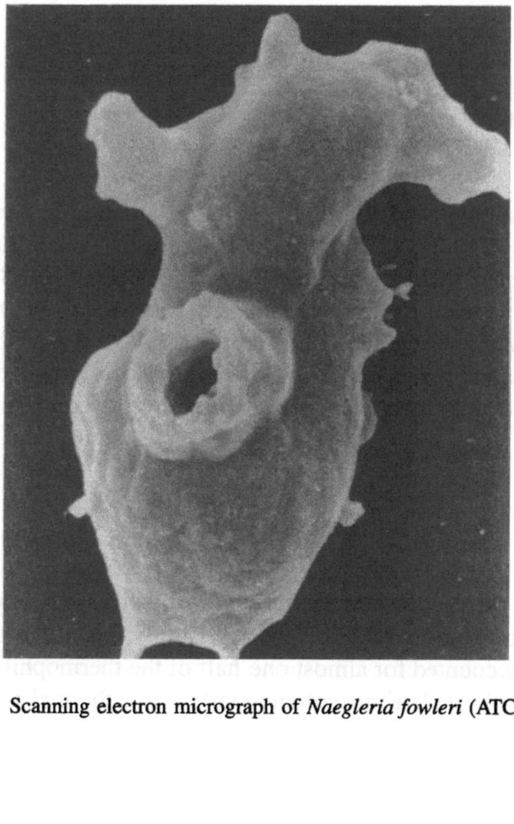

Figure 1. Scanning electron micrograph of *Naegleria fowleri* (ATCC 30863).

Table 3
Growth of *Naegleria* Species on Thermophilic Bacteria

Bacteria	Temperature	Medium	*fowleri*[a]	*australiensis*	*lovaniensis*
YT-1 (Yellow) (70°C)	35°C	Castenholz	+	+	−
	35°C	Page amoeba saline	+	+	+
	44°C	Castenholz	+	+	−
	44°C	Page amoeba saline	+	+	±
X-1 (White) (70°C)	35°C	Castenholz	+	+	−
	35°C	Page amoeba saline	+	+	+
	44°C	Castenholz	+	+	−
	44°C	Page amoeba saline	+	+	−
S-1 (White) (70°C)	35°C	Castenholz	+	+	±
	35°C	Page amoeba saline	+	+	±
	44°C	Castenholz	+	+	−
	44°C	Page amoeba saline	+	+	+
K-2 (Red) (60°C)	35°C	Castenholz	+	+	+
	35°C	Page amoeba saline	+	+	+
	44°C	Castenholz	+	+	−
	44°C	Page amoeba saline	+	+	−
Huckleberry (Red) (60°C)	35°C	Castenholz	+	+	+
	35°C	Page amoeba saline	+	+	+
	44°C	Castenholz	+	+	−
	44°C	Page amoeba saline	+	+	−
T. ruber (Red) (60°C)	35°C	Castenholz	+	+	−
	35°C	Page amoeba saline	+	+	±
	44°C	Castenholz	+	+	−
	44°C	Page amoeba saline	+	+	−
E. coli (37°C)	35°C	Castenholz	+	+	+
	44°C	Page amoeba saline	+	+	+
	44°C	Page amoeba saline	+	+	+

[a]+ = growth equivalent to control; ± = growth less than control; − = no additional growth.

from Colorado hot springs and thermally polluted streams in Indiana (Ramaley and Bitzinger, 1975, and Ramaley et al., 1975), Huckleberry [a red–pigmented strain also similar to *Thermus ruber* that was isolated from Huckleberry Hot Spring (located in the Rockefeller Parkway (Grand Teton National Park) just south of Yellowstone National Park)] and *Thermus ruber* (Loginova and Egorova, 1975). All of these *Thermus* strains served as food sources for *Naegleria* . The apparent reduced growth of *N. lovaniensis* may be due to the reduced migration of *N. lovaniensis* compared to *N. fowleri* (Thong and Ferrante, 1986).

4. DISCUSSION

We documented the presence of thermophilic amoebae in hot springs of the Yellowstone and Grand Teton National Parks. Not suprisingly, using the described methods of isolation, *Naegleria* accounted for almost one-half of the thermophilic amoebae observed. Although *Naegleria* is the isolate most often reported from thermal sources (DeJonckheere,

1977, 1979, 1981; DeJonckheere et al., 1983; Scaglia et al., 1983; Stevens et al., 1980), information on the presence of other amoebae is usually not available, so that a comparison of their relative abundance is not possible. The association of *H. vermiformis* with *Naegleria* isolates, particularly at increased temperatures of isolation, was previously reported (O'Dell and Stevens, 1973). Kyle and Noblet (1985) reported that *Hartmannella* and *Naegleria* accounted for 75–80% of the isolates from the water column of a pond and that *Acanthamoeba* represented only 8–10% of the isolates. O'Dell (1979) reported that *Acanthamoeba* is the most frequently isolated amoebae from sediment of nonthermal sources. DeJonckheere (1979) found that *Acanthamoeba* is more common than *Naegleria* in swimming pools, whereas the opposite is true of aquaria and surface waters. The absence of *Acanthamoeba* from our samples population may have been due to the extremely rapid growth of the other isolates which masked their presence. Subsequent examination of commercial swimming areas south of Grand Teton National Park and at Boiling River near Mammoth Hot Springs in Yellowstone National Park did yield *Acanthamoeba* isolates (W. D. O'Dell and P. L. Scanlan, unpublished observations).

SCGYEM axenic medium is used to differentiate pathogenic and nonpathogenic *Naegleria* isolates (DeJonckheere, 1977). One of our isolates that grew luxuriantly in SCGYEM, BHS 3-1, also showed some virulence in mice (Table 3). The lower level of virulence suggests that this isolate was *N. australiensis* (DeJonckheere et al., 1983). This particular isolate was obtained from a sample taken from the side of a commercial swimming pool outside of the park that is only "lightly" chlorinated (owner's personal comment). The pool is emptied each night and then allowed to refill for the next day. A second isolate produced the symptoms of the infection without killing the host, also a characteristic observed for *N. australiensis* (DeJonckheere, 1981). This strain, HS-2, was isolated from a sample taken from a spring (Huckleberry) that was until recently the source of water for a swimming pool. Isoenzyme analysis has verified that strains BHS 3-1 and HS-2 are *N. australiensis* (O'Dell, unpublished observations).

It should not be surprising, then, that these swimming pools and hot springs with their overflow channels provide an appropriate habitat for these thermophilic amoebae. The amoebae have a source of food (*Thermus* and other bacteria, etc.) in a complex microbial mat that makes an ideal substrate on which to graze, and an elevated temperature that selects against possible mesophilic competitors. Because potentially pathogenic *Naegleria* species also occupy this biological niche, the demonstrated presence of thermophilic amoebae suggests that by more intense sampling, isolates such as *N. fowleri* will also be found in these springs and pools.

It turns out that *Naegleria* grows quite well on *Thermus* isolates, and it is possible that Thermus may well be one of the natural hosts of *Naegleria* in lower temperature hot springs and runoff channels. One experimental advantage of using *Thermus* as a feeding layer for *Naegleria* is that it will not grow at 37°C, and thus *Thermus* cultures can be used as a transitional substrate to obtain axenic cultures free from any contamination by mesophilic bacteria such as *E. coli*.

Thus, it has been possible to demonstrate the presence of thermophilic *Naegleria* isolates from low temperature hot spring and runoff channels which park visitors and employees use for bathing (Whittley, 1995). Although pathogenic isolates of *N. fowleri* were not found in these areas, cultural conditions do exist for potential growth and infection by pathogenic Naegleria .

5. SUMMARY

Thermophilic (45°C) amoebae were found in the lower temperature, neutral to slightly alkaline thermal springs and pools in Yellowstone and Grand Teton National Parks. Sixteen of the first thirty-four Yellowstone/Grand Teton thermophilic amoebic isolates were identified as *Naegleria* genera, and *Naegleria* isolates were found in eight of the sites that yielded thermophilic amoebic isolates. No thermophilic amoebae were isolated from acidic springs. One of the sites that showed consistently high levels of *Naegleria* was the Huckleberry Hot Spring/Polecat Creek area in the north part of Grand Teton National Park. Those areas that showed high levels of *Naegleria* have now been posted with warning signs by the U.S. National Park Service and the U.S. Public Health Service (Centers for Disease Control) to reduce human exposure to *Naegleria* which can cause human primary amoebic meningoencephalitis and which, although the incidence is rare, is inevitably fatal.

Based on these studies, the U.S. Park Service (in cooperation with the U.S. Public Health Service (Center for Disease Control)) has posted informational warning signs at Boiling River (near Mammoth Hot Springs) in Yellowstone National Park and at Huckleberry Hot Springs in Grand Teton National Park where the highest levels of *Naegleria* were found. Bathers who use these areas, and especially the more isolated springs along Polecat Creek in Grand Teton National Park, often dig pools and channels to mix cold and hot water to control bathing temperature. These bathers are probably at greater risk for amoebic meningoencephalitis because as John (1982) suggested, exposure to amoebae is more likely to occur in such a disturbed environment. However, as long as bathers do not immerse their heads in the water and do not inhale water up their noses, it does not pose any immediate threat.

Although the object of the initial research reported in this chapter was to determine the potential human public health hazard from pathogenic *Naegleria* from direct culturing, it is obvious that the ecology and physiology of these organisms are receiving increasing research attention. The development of specific PCR primers by Kilvington and Beeching (1995) which allows detection at the level of a single amoeboid trophozoite or cyst will greatly increase the sensitivity of ecological studies. Studies by John and Hoppe (1990) showed that rodents are also susceptible to *N. fowleri*, and antibodies against *Naegleria* have been detected in thirteen wild mammalian species collected in southwestern Tennessee (Kollars and Wilhelm, 1996).

Thus "natural thermal areas" protected by the U.S. National Park Service, such as those of the lower temperature spring at Huckleberry Hot Spring in Grand Teton National Park (which has now been de-developed back from a commercial swimming pool to a "natural" spring by the U.S. Park Service) and the nearby springs along Polecat Creek have provided a stable natural population of thermophilic amoebae for study during a 10-year period (W. O'Dell, unpublished observation). These sites should provide a valuable resource for more detailed examination of the ecology and effect of *Naegleria* and the role of these and other thermophilic and mesophilic amoebae in the ecology of these lower temperature thermal hot springs (Scanlan, 1988).

Acknowledgments. This work was supported in part by the University Committee on Research, University of Nebraska at Omaha, and by a grant from the National Institutes of Health. The scanning electron micrograph of *Naegleria fowleri* was prepared by Pam Scanlan. The assistance of Daniel O'Dell, Alan Ramaley, and Andrew Ramaley in collecting the field samples is gratefully acknowledged. The facilities and cooperation of the personnel at the University of Wyoming, U.S. National Park Service Field Station, was invaluable to the follow-up studies at Huckleberry and Polecat Creek Hot Springs. The assistance of Han Huang in converting manuscript computer files was greatly appreciated. The cooperation and advice of numerous U.S. National Park personnel including John Varley and Robert Lindstrom of the Research Division at Yellowstone National Park and William Barmore, Research Biologist at Grand Teton National Park were especially appreciated.

REFERENCES

Boyle, J., Kingston, N., and Jolley, W. 1982. Thermophilic amoebae from Yellowstone Hot Springs. Annual Meeting, *Am. Soc. Parasit Ab.* 31.

Brock, T. D., and Freeze, H. 1969. *Thermus aquaticus*, gen. n. and sp. n., a nonsporulating extreme thermophile. *J. Bacteriol.* **98:**289–297.

Cerva, L. 1969. Amoebic meningoencephalitis: axenic culture of *Naegleria*. *Science* **163:**576.

DeJonckheere, J. 1977. Use of an axenic medium for differentiation between pathogenic and non-pathogenic *Naegleria fowleri* isolates. *Appl. Environ. Microbiol.* **33:**751–757.

DeJonckheere, J., and Van De Voorde, H. 1977. The distribution of *Naegleria fowleri* in man-made thermal waters. *Am. J. Trop. Med. Hyg.* **26:**10–15.

DeJonckheere, J. F. 1979. Occurrence of *Naegleria* and *Acanthamoeba* in aquaria. *Appl. Environ. Microbiol.* **38:**590–593.

DeJonckheere, J. 1981. *Naegleria australiensis* sp. nov., another pathogenic *Naegleria* from water. *Protistologica* **17:**423–429.

DeJonckheere, J. F., Aerts, M., and Martinez, A. J. 1983. *Naegleria australiensis*: Experimental meningoencephalitis in mice. *Trans. R. Soc. Trop. Med. Hyg.* **77:**712–716.

DeJonckheere, J. F., Pernin, P., Scaglia, M., and Michel, R. 1984. A comparative study of 14 strains of *Naegleria australiensis* demonstrates the existence of a highly virulent subspecies: *N. australiensis italica* n. ssp. *J. Protozool.* **31:**324–331.

Jensen, T., and Dubes, G. R. 1962. Cloning, titration, and differentiation of *Acanthamoeba* sp. by plating. *J. Parasitol.* **48:**280–286.

John, D. T. 1982. Primary amebic meningoencephalitis and the biology of *Naegleria fowleri*. *Annu. Rev. Microbiol.* **63:**101–123.

John, D. T., Cole, T. B., and Marciano-Cabral, F. 1984. Sucker-like structure on the pathogenic amoeba *Naegleria fowleri*. *Appl. Environ. Microbiol.* **47:**12–14.

John, D. T., and Hoppe, K. L. 1990. Susceptibility of wild mammals to infection with *Naegleria fowleri*. *J. Parasitol.* **76:**865–868.

Kilvington, S., and Beeching, J. 1995. Development of a PCR for identification of *Naegleria fowleri* from the environment. *Appl. Environ. Microbiol.* **61:**3764–3767.

Kollars, T. M., and Wilhelm, W. E. 1996. The occurrence of antibodies to *Naegleria* species in wild mammals. *J. Parasitol.* **82:**73–77.

Kyle, D. E., and Noblet, G. P. 1985. Vertical distribution of potentially pathogenic free-living amoebae in freshwater lakes. *J. Protozool.* **32:**99–105.

Loginova, L. G., and Egorova, L. A. 1975. *Thermus ruber* obligate thermophilic bacteria in thermal springs of Kamchatka. *Mikrobiologiia* **44:**661–665.

Marciano-Cabral, F., 1988. Biology of *Naegleria* spp. *Microbiol. Rev.* **52:**114–133.

Martinez, A. J., and Visvesvara, G. S. 1997. Free-living, amphizoic and opportunistic amebas. *Brain Pathol.* **7:**583–598.

Mitchell, L. G., Mutchmor, J. A., and Dolphin, W. D. 1988. *Zoology*. Menlo Park, CA: Benjamin/Cummings.

O'Dell, W. D., and Stevens, A. R. 1973. Quantitative growth in *Naegleria* in axenic culture. *Appl. Microbiol.* **25**:621–627.

O'Dell, W. D. 1979. Isolation, enumeration and identification of amoebae from a Nebraska lake. *J. Protozool.* **26**:265–269.

Page, F. C. 1967. Taxonomic criteria for limax amoebae, with description of 3 new species of *Hartmannella* and 3 of *Vahlkampfia*. *J. Bacteriol.* **98**:289–297.

Ramaley, R. F., and Hixson, J. 1970. Isolation of a nonpigmented, thermophilic bacterium similar to *Thermus aquaticus*. *J. Bacteriol.* **103**:527–528.

Ramaley, R. F., and Bitzinger, K. 1975. Types and distribution of obligate thermophilic bacteria in man-made and natural thermal gradients. *Appl. Microbiol.* **30**:152–155.

Ramaley, R. F., Bitzinger, K., Carroll, R. M., and Wilson, R. B. 1975. Isolation of a new pink thermophilic bacterium (K-2 Isolate). *Int. J. Syst. Bacteriol.* **25**:357–364.

Scaglia, M., Storsselli, M., Grazioli, V., Gatti, S., Bernuzzi, A. M. and DeJonckheere, J. F. 1983. Isolation and identification of pathogenic *Naegleria australiensis* (Amoebida, Vahlkampfidae) from a spa in Northern Italy. *Appl. Environ. Microbiol.* **46**:1282–1285.

Scanlan, P. L. 1988. Application of the API ZYM system in the identification of *Naegleria* isolates. M.S. Thesis, University of Nebraska at Omaha.

Stevens, A. R., DeJonckheere, J., and Willaert, E. 1980. *Naegleria lovaniensis* new species: Isolation and identification of six thermophilic strains of a new species found in association with *Naegleria fowleri*. *Int. J. Parasit.* **10**:51–64.

Thong, Y. H., and Ferrante, A. 1986. Migration pattern of pathogenic and nonpathogenic *Nagleria* spp. *Infect. Immun.* **51**:177–180.

Whittley, L. H. 1995. *Death in Yellowstone*. Boulder, CO: Roberts Rinehart.

Willaert, E. 1971. Isolement et culture in vitro des amibes du genre *Naegleria*. *Ann. Soc. Belge Med. Trop.* **51**: 701–708.

5

Examining Bacterial Population Diversity within the Octopus Spring Microbial Mat Community

M. J. Ferris, S. C. Nold, C. M. Santegoeds, and D. M. Ward

1. INTRODUCTION

The importance of preserving biodiversity has gained considerable attention as the possibility of destroying our remaining natural ecosystems has been realized (Schulze and Mooney, 1993). Consequently questions about how the loss of vast, species-rich habitats such as rain forests might influence global scale processes are receiving serious consideration. A more pragmatic view holds that a reduction in species diversity might result in the loss of valuable commodities, such as natural compounds with medicinal properties. Certainly bacteria influence processes on a global scale and produce many of humanity's most beneficial and commercially valuable products, yet sequence-based population surveys have proven that only a small minority of bacteria have been studied. Some recent findings underscore this point. In 1993, it was discovered that macroscopic symbionts of surgeonfish are actually bacterial cells (Ahern, 1993; Angert et al., 1993), designated *Epulopiscium fishelsoni*. This bacterium is the largest prokaryote known, and its existence has forced a reevaluation of the limit of prokaryotic cell size. In the same year, new bacteria that perform a novel type of photosynthesis, phototropic iron oxidation, were discovered

M. J. Ferris, S. C. Nold, C. M. Santegoeds, and D. M. Ward • Department of Microbiology, Montana State University, Bozeman, Montana 59717.

Thermophiles: Biodiversity, Ecology, and Evolution, edited by Reysenbach *et al.* Kluwer Academic / Plenum Publishers, New York, 2001.

(Widdel et al., 1993; Ehrenreich and Widdel, 1994), presenting a possible new explanation for the origin of banded iron formations. In 1994, an entirely new branch of Archaea was discovered. Termed the Korarchaeota, these bacteria represent the most deeply divergent archaeal group known (Barns et al., 1994, 1996). With such striking diversity now revealed, it seems certain that new insights into nutrient cycling and bacterial metabolites remain to be discovered.

A major factor contributing to our limited understanding of bacterial diversity and ecology has been about a century of reliance upon techniques, such as microscopy and cultivation, that fail to reveal most of the microorganisms in nature (Ward et al., 1992). The development of molecular sequence-based approaches to bacterial population surveys has provided a novel alternative to traditional enrichment culture or microscopic analyses of natural samples. In many cases, reasonable identifications and phylogenetic affiliations can now be made based on the retrieval of nucleotide sequences (Ward et al., 1992; Olsen and Woese, 1993; Woese, 1987, 1994; Olsen et al., 1986; Pace et al., 1986; Amann et al., 1994). Because sequence information can be extracted directly from the environment, the requirement for cultivation and microscopy is to some extent circumvented. Microbial ecologists can now survey bacterial populations in a fashion that more nearly parallels surveys of macroscopic organisms.

We have applied both traditional and molecular techniques to gain a more complete understanding of the microbial diversity of the cyanobacterial mat community in Octopus Spring, a thermal pool in the Lower Geyser Basin of Yellowstone National Park (Ward et al., 1987, 1992, 1994, Chapter 12, this volume). A thermal habitat facilitates our studies by providing a more constant environment and limiting the biota to prokaryotic forms. As in any study of community ecology, a fundamental aspect of our research has been to identify as many of the populations within the mat as possible. We define populations on the basis of 16S rRNA sequences, realizing that this approach, as with all methods used to detect or classify bacteria, is not without limitations. Uncertainties associated with molecular techniques, such as primer bias, chimera formation during PCR, or interoperon sequence variation (Kopczynski et al., 1994; Malvagnam and Dennis, 1992; Nubel et al., 1996; Robison-Cox et al., 1995; Rainey et al., 1994, 1996; Reysenbach et al., 1992) and the conservative nature of the 16S rRNA gene may underestimate actual diversity (Ward et al., 1992). Here, we present an updated list of the microbial populations thus far catalogued and describe some of the more recent methods we have used to detect them.

Initial surveys of the Octopus Spring mat community revealed that none of the 16S rRNA sequences of any cultivated isolate, even those obtained from Octopus Spring or similar thermal environments, matched those retrieved from the mat by direct cloning (Ward et al., 1990a, b; Weller et al., 1991, 1992). A similar pattern of disparity has been demonstrated in terrestrial, marine, and other environments (DeLong, 1992; Fuhrman et al., 1993; Giovannoni et al., 1990; Liesack and Stackebrandt, 1992; Schmidt et al., 1991; Ward et al., 1994; Choi et al., 1994; Suzuki et al., 1997). Oligonucleotide probe analyses of Octopus Spring cyanobacterial enrichments indicated that some of the populations detected by cloning could be selectively enriched if inocula were sufficiently diluted before incubation (Ferris et al., 1996b). We subsequently used extincting dilution enrichment techniques to obtain axenic isolates of some bacterial populations whose 16S rRNA sequences were detected by cloning. More recently, we have incorporated denaturing gradient gel electrophoresis (DGGE) analysis of PCR-amplified 16S rRNA gene segments into our survey of the populations inhabiting the Octopus Spring mat (Ferris et al., 1996a). This has expanded the list of populations detected in the mat, but perhaps more importantly, the combined results of all our studies have stimulated a reevaluation of our view of diversity within this system and the ecological and evolutionary basis for its existence (Ward et al., 1994, 1997, in press; Ferris et al., 1996a; Nold and Ward, 1995).

2. OCTOPUS SPRING MAT CYANOBACTERIAL DIVERSITY AS REVEALED BY MICROSCOPY, CULTIVATION, PROBING, CLONING, AND SEQUENCING

Examined macroscopically, a vertical section through the Octopus Spring mat reveals a top green layer approximately 1.0 mm thick. Microscopic examination of this layer reveals numerous rod-shaped, autofluorescing unicells previously believed to be the cyanobacterium *Synechococcus lividus* (Ward et al., 1989). This single species designation was based primarily on morphology, the observation that these cell types are commonly found in thermal spring mats under conditions similar to those in the Octopus Spring mat (low sulfide, pH approximately 8.5, temperature range approximately 55 to 65° C) and that cells that have this morphology are readily enriched under standard photoautotrophic conditions. Furthermore, all cultivated thermophilic *Synechococcus* strains have similar mole % G+C ratios (Castenholz and Waterbury, 1989; Castenholz, 1981).

Results of analyses of 16S rDNA sequences cloned directly from Octopus Spring mat samples revealed the presence of five cyanobacterial populations (Table 1, O.S. type-A, -B, -I, -J and -P). Curiously, none of these sequences matched that of *S. lividus* (Ward et al., 1994; Weller et al., 1991). Subsequent studies were conducted using radiolabeled oligonucleotide probes specific for individual Octopus Spring cyanobacterial populations and the *S. lividus* sequence. These confirmed the presence of type-A-like, type-B-like and type-J-like populations (Table 1) and suggested that *S. lividus* might be present, but at levels much lower than those of the cyanobacterial populations detected by cloning (Ruff-Roberts et al., 1994; Ferris et al., 1996b). The probe studies also suggested that *S. lividus* grew well under enrichment culture conditions and that some of the cyanobacterial populations detected by cloning could be selectively enriched by simply diluting homogenized mat samples before inoculating enrichment media (Ferris et al., 1996b). Using this approach, we obtained two axenic cyanobacterial isolates from near-extincting dilution enrichments (Table 1). One of these, isolate B-10, obtained from an enrichment inoculated with approximately 25 *Synechococcus*-shaped cells, does contain a 16S rRNA sequence that matches a clone retrieved directly from the mat, clone type-P. However, the other isolate, C9, obtained from an enrichment inoculated with approximately 250 *Synechococcus*-shaped cells, contains a novel sequence that has not yet been detected by molecular methods. Initial direct counts of *Synechococcus*-like cells ($\sim 1 \times 10^{10}$ cells/mL) indicate that both isolates were likely to have been among the predominant cyanobacterial populations in the samples from which they originated. Conversely, two isolates from low dilution enrichments, isolates B1 and C1, have sequences that are identical to each other and to that of *S. lividus* Y-7c-s, a strain originally isolated from another Yellowstone hot spring (Kallas and Castenholz, 1982). This result is consistent with the probe data and suggests that the *S. lividus* Y-7c-s strain is particularly competitive under laboratory conditions and might mask the presence

Table 1
Populations Defined by 16S rRNA Sequences Retrieved from the Octopus Spring Mat by Cultivation, Cloning, DGGE Band Sequencing of Mat DNA or DGGE Band Sequencing of DNA Extracted from Enrichment Cultures Using Mat Samples as Inocula

Phylogenetic type population	Isolated culture	16S rRNA clone	DGGE mat	DGGE enrichment	16S rRNA probe	References[a]
Cyanobacteria						
Isolate C1 (*Synechococcus lividus* Y-7c-s)	×				×	1, 2
O.S.type-A		×	×		×	3, 4
O.S.type-A′			×		×[b]	5
O.S.type-A″			×			6
O.S. type-A‴			×			7
O.S.type-B		×	×		×	3,4
O.S.type-B′	×[c]		×		×[b]	5
Isolate C9	×					1
O.S.type-I		×				4
O.S.type-J		×				4
O.S.type-P (Isolate B10)	×	×				1
Green nonsulfur bacteria-like						
Chloroflexus aurantiacus Y-400-fl	×				×	
Thermomicrobium roseum	×[d]					9–11
O.S.type-C		×	×		×	3–5
O.S.type-C′			×			5
O.S.type-C″			×			6
OS-V-L-20		×				4
Env.OS_ace3				×		12
Env.OS_ace4				×		12
Env.OS_ace5				×		12
Green sulfur bacteria-like						
O.S.type-E		×				3
O.S.type-E′			×			5
O.S.type-E″			×			5
O.S.type-M		×				3
O.S.-III-9		×				13
Planctomyces						
Isophera pallida IS1B	×[e]					14, 15
Thermus sp.						
Thermus sp. ac-1 (T. aquaticus-like)	×			×		12, 16, 17
Thermus sp. ac-2 (T. ruber-like)	×			×		12, 16, 17
Thermus sp. ac-7 (T. aquaticus-like)	×					17
Thermus sp. ac-7′ (T. aquaticus-like)				×		12
Thermus sp. ac-17 (T. ruber-like)	×					17
Thermus sp. O.S. Ramaley-4	×					17
Proteobacteria						
Alpha subdivision						
O.S.type-O		×				13
Env.OSace_2				×		12
Beta subdivision						
O.S.type-G		×				3
O.S.type-R		×				18
O.S.type-N (Isolate ac-15)	×	×		×		12, 18

Table 1 (*Continued*)

Phylogenetic type population	Isolated culture	16S rRNA clone	DGGE mat	DGGE enrichment	16S rRNA probe	References[a]
Proteobacteria (*cont.*)						
Beta subdivision (*cont.*)						
O.S.type-N′				×		12
Isolate ac_16	×			×		12
Env.OS ace_7				×		12
Gamma subdivision						
Env.OSace_8				×		12
Gram-positive bacteria						
Thermoanaerobium brockii HTD4	×[f]					19
Thermobacteroides acetoethylicus HTB2/W	×					20
Thermoanaerobacter ethanolicus JW200	×					21
Clostridium thermohydrosulfuricum 39E	×					22
Clostridium thermosulfurogens 4B	×					23
Clostridium thermoautotrophicum JW701	×					24
Env.OS_ace1				×		12
Env.OS_ace6				×		12
Thermodesulfobacterium						
Thermodesulfobacterium commune YSRA	×[g]					25
Spirochetes						
O.S.type-H		×				3
OS-V-L-7		×				4
O.S.type-K		×				4
Uncertain affiliation						
O.S.type-D		×				3
O.S.type-F		×				3
OP-I-2		×				26
O.S.type-L		×		×		26
O.S.type-Q				×		5
Archaea						
Methanobacterium thermoautotrophicum Δ H	×[h]					20, 27, 28

[a]References: 1. Ferris et al., 1996b. 2. Kallas and Castenholz, 1982. 3. Ward et al., 1990b. 4. Weller et al., 1992. 5. Ferris et al., 1996a. 6. Ferris and Ward, 1997. 7. Ferris et al., 1997. 8. Pierson and Castenholz, 1974. 9. Jackson et al., 1973. 10. Zeng et al., 1992a. 11. Zeng et al., 1992b. 12. Santegoeds et al., 1996. 13. Ward et al., 1992. 14. Doemel and Brock, 1977. 15. Giovannoni et al., 1987. 16. Brock and Freeze, 1969. 17. Nold and Ward, 1995. 18. Nold et al., 1996. 19. Zeikus et al., 1979. 20. Zeikus et al., 1980. 21. Wiegel and Ljungdahl, 1981. 22. Wiegel et al., 1979. 23. Schink and Zeikus, 1983. 24. Wiegel et al., 1981. 25. Zeikus et al., 1983. 26. Ward et al., 1994. 27. Sandbeck and Ward, 1982. 28. Zeikus and Wolfe, 1972.

[b]Probes targeting O.D.type-A and O.S.type-B cyanobacterial populations will also detect O.S.type-A′ and O.S.type-B′ (16).

[c]Detected first by DGGE, then by reanalysis of cloned sequences.

[d]Cultivated from Toadstool Spring, but presumed to inhabit Octopus Spring because distinctive diol lipids are abundant in mat.

[e]Cultivated from Kah-nee-tah Spring, but presumed to inhabit Octopus Spring because of the presence of organisms with similar morphology.

[f]Cultivated from Washburn Spring, but Octopus Spring strains are known.

[g]Cultivated from Ink Pot Spring, but Octopus Spring strains are known.

[h]Cultivated from sewage sludge, but Octopus Spring strains are known.

of other more numerically relevant strains during the process of selective enrichment. To evaluate this further, we obtained partial 16S rDNA sequence data for all four available thermophilic *Synechococcus* strains in culture collections. We found that they too match *S. lividus* strain Y-7c-s, lending more credence to the hypothesis that *S. lividus* is particularly well adapted to the environment provided by enrichment approaches (Ferris et al., 1996b). All of the isolates have a common *Synechococcus*-like morphology and are microscopically indistinguishable from each other.

3. STANDARDIZATION OF METHODOLOGY, ENVIRONMENTALLY MEANINGFUL SAMPLING POINTS, AND INCREASED SAMPLE THROUGHPUT ARE NECESSARY TO UNDERSTAND OCTOPUS SPRING POPULATION DYNAMICS

Through a variety of approaches, much has been learned about the diversity of microbes in the Octopus Spring mat community; however, few correlations can be made between populations detected and relevant environmental parameters that might control their distributions because the multifaceted nature of the experiments that have generated the current population data do not exclude methodological and spatiotemporal sampling variations. To better understand population dynamics within the Octopus Spring mat, a more robust set of population distribution data will be necessary (Ward et al., in press). We feel that these data should be obtained using a single technique on replicate samples collected during ecologically relevant intervals. To accomplish this, we have employed DGGE as the principal approach to population analysis (Ferris et al., 1996a). With DGGE, partial 16S rRNA sequences are readily obtained from PCR-amplified 16S rDNA bands excised from denaturing gradient gels. We have begun to collect data on population distributions along natural thermal gradients (Ferris and Ward, 1997) and to examine the effects of disturbances on population succession such as those that might be caused by hailstorms or trampling (Ferris et al., 1997). We have also examined how seasonal and diurnal changes in temperature and light influence population distributions (Ferris and Ward, 1997; Ramsing et al., 2000).

DGGE can be used to separate mixtures of PCR-amplified 16S rDNA segments based on their sequence differences. The PCR products are electrophoresed through a polyacrylamide gel containing a linearly increasing concentration gradient of denaturants. Figure 1 illustrates the DGGE separation of five unique, PCR-amplified, Octopus Spring cyanobacterial 16S rDNA segments. The five samples were combined and loaded into each lane of the gel at regular time intervals. Initially all of the PCR products migrate as a single band because they are the same size. As migration proceeds, the sequences encounter increasing concentrations of denaturants. Eventually, based on their stability, a function of their sequence, each fragment reaches a unique position in the gradient where it begins to denature. This partial denaturation lowers the electrophoretic mobility of the fragment dramatically and brings its migration to halt. After 195 min, all of the PCR products separated and the banding pattern remained essentially the same for the duration of the experiment. More detailed descriptions of the DGGE technique have been presented

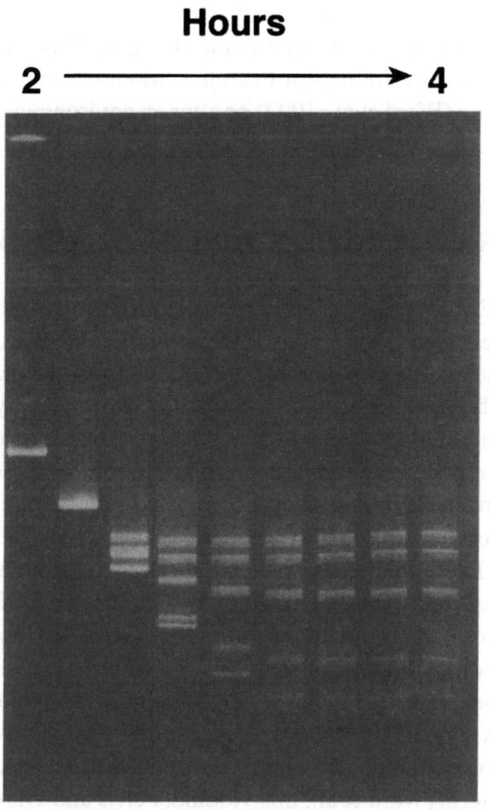

Figure 1. Denaturing gradient gel demonstrating the separation of five PCR-amplified Octopus Spring cyano-bacterial 16S rDNA gene segments with time of electrophoresis.

elsewhere (Muyzer et al., 1993; Myers et al., 1987; Abrams and Stanton, 1992; Muyzer and De Waal, 1994).

Using DGGE, only partial 16S rDNA segments are typically evaluated. Obviously, this limits the amount of phylogenetically useful nucleotide sequential information available. However, the advantage is that multiple populations can be detected simultaneously, which greatly increases the number of samples that can be analyzed. This allows greater amounts of ecologically useful information to be obtained relatively quickly. In contrast, oligonucleotide probe studies limit detection to known populations. Surveying multiple populations makes DGGE more applicable for the comprehensive population analyses we have undertaken. New populations, represented by new bands in the gradient, can be selected for sequencing (Ferris et al., 1996a; Muyzer and De Waal, 1994; Muyzer et al., 1995; Wawer and Muyzer, 1995). Even faint bands, which would presumably represent a minority of clones in a PCR-generated library, can be identified and sequenced directly. This eliminates possible bias and redundancy associated with randomly sequencing 16S rDNA clones (Rainey et al., 1994; Farrelly et al., 1995). In addition, DGGE banding profiles can provide rapid information about population changes. This can be useful when numerous sampling points are desirable, for example, when monitoring population changes during enrichment culture (Ward et al., 1997) or after an environmental perturbation (Ferris et al., 1997).

4. DGGE ANALYSIS OF OCTOPUS SPRING MAT SAMPLES

We applied DGGE analysis to DNA extracted from regions of the Octopus Spring mat defined by temperature range (Ferris et al., 1996a; Ferris and Ward, 1997). Results from these experiments revealed that populations within regions of the mat defined by the same temperature range are homogeneous, whereas distinct heterogeneity is exhibited between populations that inhabit areas of different temperature range. Sequence analysis of individual bands revealed a pattern of temperature distribution for a closely related group of Octopus Spring cyanobacterial populations (designated type-A-like and type-B-like) consistent with earlier probe analyses (Ruff-Roberts et al., 1994; Ferris and Ward, 1997). However, because the DGGE analysis resolved additional A-like and B-like populations (designated types-A′, -A″ and type-B′; Table 1), that cannot be distinguished by using the type-A and type-B probes, respectively, a more complete description of the temperature distributions of these populations was achieved. DGGE indicates adaptations from low to high temperature in the order B, B′, A, A′, A″. The high percentage similarity (>95%) between the 16S rDNA sequences of the type-A-like and type-B-like populations suggests that temperature-driven population divergence is a comparatively recent evolutionary event among these cyanobacteria (Ward et al., this volume; Ferris and Ward, 1997). The detection of a novel cyanobacterial population after disturbing the mat (A‴, Table 1) suggests that some species may be adapted as colonists.

As indicated before, DGGE analysis of the Octopus Spring mat confirmed the presence of previously detected cyanobacterial populations and revealed new, closely related populations. Similar results have been observed for other groups of Octopus Spring bacteria. The known green nonsulfur bacteria-like population, type-C and two newly discovered, related populations, types -C′ and C″, have been detected and show temperature distribution differences (Ferris and Ward, 1997). Two new green sulfur bacteria-like populations (types-E′ and -E″), closely related to the cloned type-E sequence, have also been detected (Table 1).

It is worth noting that the utility of DGGE in separating sequences that differ by as little as a single nucleotide (Myers et al., 1987) permitted us to resolve populations that were ambiguous by sequencing 16S rRNA clones. For example, cyanobacterial population type-B′ was actually detected among clones in our Octopus Spring mat 16S rRNA libraries, but due to the possibility of sequencing error, the type-B and type-B′ sequences were originally combined and designated cyanobacterial population type-B. Because the DGGE analyses facilitated repeated samplings that separated the type-B and -B′ sequences in the denaturing gradient gel and demonstrated that they occurred independently in different temperature regions of the mat, we were able to determine that the sequences represented two independent populations (Ferris et al., 1996a).

5. DGGE ANALYSES OF AEROBIC CHEMOORGANOTROPHIC ENRICHMENT CULTURES DEMONSTRATES THE INCONGRUENCE AMONG POPULATIONS WITHIN NATURAL MICROBIAL COMMUNITIES AND SELECTIVE ENRICHMENT CULTURES

We used dilution enrichment culture to obtain diverse *Thermus* and other aerobic chemoorganotrophic isolates. At present, only one of these isolates has a 16S rDNA sequence that matches that of a clone retrieved directly from the mat, Octopus Spring type-N (Table 1) (Nold and Ward, 1995; Nold et al., 1996). This result is typical. Studies of various natural communities have repeatedly demonstrated a nearly total lack of correspondence between the bacterial populations that have been cultivated and those that have been detected by molecular methods (Barns et al., 1994; Britschgi and Giovannoni, 1991; Choi et al., 1994; Giovannoni et al., 1990; Schmidt et al., 1991; Stackebrandt et al., 1993; Ward et al., 1990a, b; Weller et al., 1992; Liesack and Stackebrandt, 1992; Suzuki et al., 1997). Thus, present cultivated species provide a poor basis from which to draw inferences about the ecology of native microbial communities. One of the long-term goals of our research is to explain this disparity and to obtain the as yet uncultivated populations in pure culture. Pure cultures are essential for physiological studies that can link phenotypes with 16S rRNA sequences retrieved directly from environmental samples.

We used DGGE analyses to evaluate the influence of selective enrichment culture on aerobic chemoorganotrophic Octopus Spring mat populations to gain insight into the disparity between populations retrieved as isolated cultures and those detected directly by molecular methods (Santegoeds et al., 1996). DGGE profiles show the effect of enrichment culture and dilution of inocula on Octopus Spring mat populations (Fig. 2). Lane M shows a profile obtained from direct analysis of an Octopus Spring mat sample taken from an approximately 55° C site. The remaining lanes are profiles generated from 21-day enrichment cultures originally inoculated with the same homogenized, 10-fold serially diluted, Octopus Spring mat sample. Few of the bands align in the DGGE gradient, indicating that population structure changes dramatically from the initial mat sample, as well as across all dilutions. We sequenced fourteen unique bands from DGGE analyses of aerobic chemoorganotrophic enrichments (Table 1). Three of these sequentially-defined populations match those of isolates previously cultivated from the Octopus Spring mat, *Thermus* populations ac-1, 2 and β-proteobacterial population ac-15. Two of these match sequences retrieved directly from the mat by cloning (Octopus Spring population type-N and -L). The rest have not been previously detected by any method. These include three *Chloroflexus aurantiacus*-like populations, a *T. aquaticus*-like population, α-, β-, and γ-proteobacterial populations, and two low G+C subdivision gram-positive bacterial populations (Table 1).

6. SUMMARY

Fifty-nine different bacterial populations have been detected in the Octopus Spring microbial mat community. About half of these have been detected directly in the mat by 16S rRNA-based methods. The others were detected in enrichment broths or from cultivated isolates. Presently, only three directly retrieved 16S rRNA sequences match those retrieved

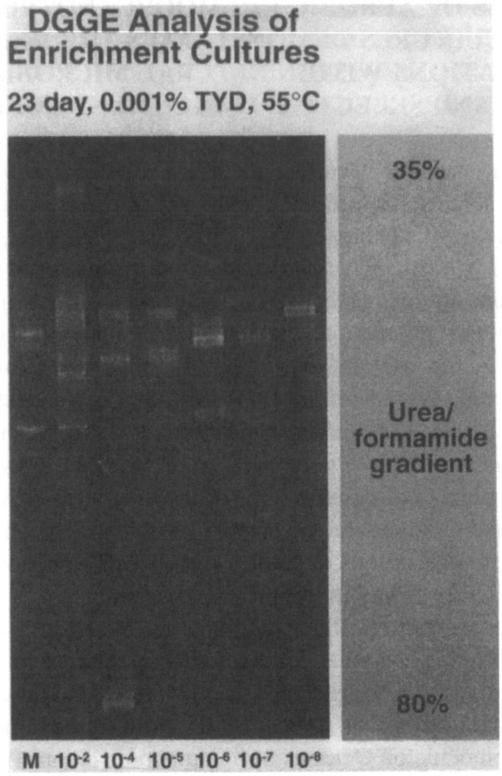

Figure 2. Denaturing gradient gel electrophoresis profiles of DNA extracted from 23-day broth-enrichment cultures of 0.001% tryptone and yeast extract in Castenholz medium D inoculated with 10-fold serially diluted (as indicated) Octopus Spring 55° C mat (M), whose DGGE profile is also shown.

indirectly. We are beginning to cultivate some of the populations revealed by 16S rRNA-based methods, but successes are rare.

Cultivation and molecular studies have been used successfully to detect microorganisms in the Octopus Spring mat. However, the diverse nature of these investigations has yielded a more or less random sampling of Octopus Spring 16S rRNA sequence types since no concerted attempt was made to collect samples in an ecologically coherent fashion. Therefore, we can only speculate as to the reasons for the discrepancy between populations detected by cultivation and those detected using 16S rRNA-based methods. It is likely that a number of factors contribute to the observed disparity.

Performing an unbiased sampling of bacterial populations in nature is difficult. Any method used to retrieve bacteria or their nucleic acids involves some degree of selectivity. Cultivation requires that adequate growth conditions be provided. Although enrichment cultures are usually designed to resemble an organism's perceived niche, they most likely do a poor job of mimicking the natural environment. Complicating this approach is the potential of limitless microscale environments within which fastidious microorganisms may have evolved and adapted. Obviously, it is impossible to duplicate each of these accurately in an enrichment culture. Thus cultivation and plating media biases are almost certainly contributing to our inability to retrieve mat populations that have been detected by 16S rRNA-based methods.

Molecular methods based on the detection of 16S rDNA sequences may also inaccurately sample bacterial populations in a microbial community. Sequencing errors associated with PCR amplification sources of error include, chimera formation and primer bias. Varying template or denaturant concentrations may also skew the outcome of PCR-based population surveys. Variations in 16S rRNA gene copy number and microheterogeneity within individual cells must also be taken into account. Finally, cloning and cell lysis methods can also introduce bias.

We are presently forced to compare Octopus Spring populations collected by multiple investigators at various times from different locations in the mat. Therefore, spatiotemporal variations in mat populations might account for some of the observed differences in populations detected. DGGE profiles indicate that populations vary between regions of the mat defined by temperature. We have begun to use this method to investigate temporal variations, as well as successional population changes, that might occur following physical disturbances to the mat, such as those caused by hailstorms or trampling by bison. Initial results indicate that PCR-amplifiable populations remained stable throughout a seasonal cycle and that disturbances do not fully account for the differences between populations detected routinely by PCR-based cloning and DGGE methods and those detected infrequently by cloning or by standard enrichment culture techniques.

The results of cultivation and molecular studies on the Octopus Spring mat have provided insights and generated interesting new questions about bacterial diversity in thermal communities. New results have forced us to continually rethink and refine our model of this system. Our intent is to focus on obtaining larger amounts of ecologically meaningful population data to establish some degree of order or predictability in the populations that inhabit this intriguing microbial mat community.

ACKNOWLEDGMENTS. We thank the U.S. National Park Service and particularly Bob Lindstrom for their cooperation and support for research in Yellowstone National Park. We also thank Anna Louise Reysenbach and all those involved in making the Yellowstone Thermophile Conference such a success. This work was supported by grants from the National Science Foundation (BSR-9209677), the National Aeronautic and Space Administration (NAGW-2764), a scholarship to C.M. Santegoeds from the Rotary Foundation, and gifts from Stratagene Corp.

REFERENCES

Abrams, E. S., and Stanton, V. P. 1992. Use of denaturing gradient gel electrophoresis to study conformational transitions in nucleic acid. *Methods Enzymol.* **212:**71–104.

Ahern, H. 1993. A big bacterium-oxymoron of the microbial world. *ASM News* **59:**519–521.

Amann, R., Ludwig, W., and Schleifer, K. H. 1994. Identification of uncultured bacteria: A challenging talk for molecular taxonomists. *ASM News* **60:**360–361.

Angert, E. R., Clements, K. D., and Pace, N. R. 1993. The largest bacterium. *Nature* **362:**239–241.

Barns, S. M., Delwiche, C. F., Palmer, J. D., and Pace, N. R. 1996. Perspectives on archaeal diversity, thermophily and monophyly from environmental rRNA sequences. *Proc. Natl. Acad. Sci. USA* **93:**9188–9193.

Barns, S. M., Fundyga, R. E., Jeffries, M. W., and Pace, N. R. 1994. Remarkable archaeal diversity detected in a Yellowstone National Park hot spring environment. *Proc. Natl. Acad. Sci. USA* **91:**1609–1613.

Ben-Basset, A., and Zeikus, J. G. 1981. *Thermobacteroides acetoethylicus* gen. nov. and spec. nov., a new chemoorganotrophic anaerobic, thermophilic bacterium. *Arch. Microbiol.* **128:**365–370.

Britschgi, T. B., and Giovannoni, S. J. 1991. Phlyogenetic analysis of a natural marine bacterioplankton population by rRNA gene cloning and sequencing. *Appl. Environ. Microbiol.* **57:**1707–1713.

Brock, T. D., and Freeze, H. 1969. *Thermus aquaticus* gen. n. and sp. n., a nonsporulating extreme thermophile. *J. Bacteriol.* **98:**289–297.

Castenholz, R. W. 1981. Isolation and cultivation of thermophilic cyanobacteria. In Starr, M. P., Stolp, H., Truper, H. G., Balows, A., and Schlegel, H. G. (eds.), *The prokaryotes* (pp. 236–246). Berlin: Springer-Verlag.

Castenholz, R. W., and Waterbury, J. B. 1989. Oxygenic photosynthetic bacteria. In Staley, J. T., Bryant, M. P., Pfennig, N., and Holt, J. G. (eds.) *Bergey's manual of systematic bacteriology, Vol. 3* (pp. 1710–1806). Baltimore: Williams & Wilkins.

Choi, B. K., Paster, B. J., Dewhirst, F. E., and Gobel, U. B. 1994. Diversity of cultivable and uncultivable oral spirochetes from a patient with severe destructive periodontitis. *Infect. Immun.* **62:**1889–1895.

DeLong, E. F. 1992. Archaea in coastal marine environments. *Proc. Natl. Acad. Sci. USA* **89:**5685–5689.

Doemel, W. N., and Brock, T. D. 1977. Structure, growth, and decomposition of laminated algal-bacterial mats in alkaline hot springs. *Appl. Environ. Microbiol.* **34:**433–452.

Ehrenreich, A., and Widdel, F. 1994. Anaerobic oxidation of ferrous iron by purple bacteria, a new type of phototrophic metabolism. *Appl. Environ. Microbiol.* **60:**4517–4526.

Farrelly, V., Rainey, F. A., and Stackebrandt, E. 1995. Effect of genome size and rRNA gene copy number on PCR amplification of 16S rRNA genes from a mixture of bacterial species. *Appl. Environ. Microbiol.* **61:** 2798–2801.

Ferris, M. J., Muyzer, G., and Ward, D. M. 1996a. Denaturing gradient gel electrophoresis profiles of 16S rRNA-defined populations inhabiting a hot spring microbial mat community. *Appl. Environ. Microbiol.* **62:**340–346.

Ferris, M. J., Nold, S. C., Revsbech, N. P., and Ward, D. M. 1997. Population structure and physiological changes within a hot spring microbial mat community following disturbance. *Appl. Environ. Microbiol.* **63:**1367–1374.

Ferris, M. J., Ruff-Roberts, A. L., Kopczynski, E. D., Bateson, M. M., and Ward, D. M. 1996b. Enrichment culture and microscopy conceal diverse thermophilic *Synechococcus* populations in a single hot spring microbial mat habitat. *Appl. Environ. Microbiol.* **62:**1045–1050.

Ferris, M. J., and Ward, D. M. 1997. Seasonal distributions of dominant 16S rRNA-defined populations in a hot spring microbial mat examined by denaturing gradient gel electrophoresis. *Appl. Environ. Microbiol.* **63:** 1375–1381.

Fuhrman, J. A., McCallum, K., and Davis, A. A. 1993. Phylogenetic diversity of subsurface marine microbial communities from the Atlantic and Pacific Oceans. *Appl. Environ. Microbiol.* **59:**1294–1302.

Giovannoni, S. J., Britschgi, T. B., Moyer, C. L., and Field, K. G. 1990. Genetic diversity in Sargasso Sea bacterioplankton. *Nature* **345:**60–63.

Giovannoni, S. J., Schabtack, E., and Castenholz, R. W. 1987. *Isosphaera pallida*, gen. and sp. nov., a gliding, budding eubacterium from hot springs. *Arch. Microbiol.* **147:**276–284.

Jackson, T. J., Ramaley, R. F., and Meinschein, W. G. 1973. *Thermomicrobium*, a new genus of extremely thermophilic bacteria. *Int. J. Syst. Bacteriol.* **23:**28–36.

Kallas, T., and Castenholz, R. W. 1982. Initial pH and ATP-ADP pools in the cyanobacterium *Synechococcus* sp. during exposure to growth-inhibiting low pH. *J. Bacteriol.* **149:**229–236.

Kopczynski, E. D., Bateson, M. M., and Ward, D. M. 1994. Recognition of chimeric small-subunit ribosomal DNAs composed of genes from uncultivated microorganisms. *Appl. Environ. Microbiol.* **60:**746–748.

Liesack, W., and Stackebrandt, E. 1992. Occurrence of novel groups of the domain bacteria as revealed by analysis of genetic material isolated from an Australian terrestrial environment. *J. Bacteriol.* **174:**5072–5078.

Malvaganam, S., and Dennis, P. P. 1992. Sequence heterogeneity between the two genes encoding 16S rRNA from the halophilic Archaebacterium *Haloarcula marismortui*. *Genetics* **130:**399–410.

Martin, M. T., Sato, M. I. Z., Tiedje, J. M., Hagler, L. C. N., Dobereiner, J., and Sanchez, P. S. (eds.) 1997. Progress in Microbial Ecology. *Brazilian Society for Microbiology/International Committee on Microbial Ecology.* São Paulo, Brazil. 147–153.

Muyzer, G., and De Waal, E. C. 1994. Determination of the genetic diversity of microbial communities using

DGGE analysis of PCR-amplified 16S rDNA. In Stal, L. J., and Caumette, P. (eds.) *Microbial mats: Structure, development and environmental significance* (pp. 207–214). Berlin: Springer-Verlag.

Muyzer, G., De Waal, E. C., and Uitterlinden, A. G. 1993. Profiling of complex microbial populations by denaturing gradient gel electrophoresis analysis of polymerase chain reaction-amplified genes coding for 16S rRNA. *Appl. Environ. Microbiol.* **59:**695–700.

Muyzer, G., Teske, A., Wirsen, C. O., and Jannasch, H. W. 1995. Phylogenetic relationships of *Thiomicrospira* species and their identification in deep-sea hydrothermal vent samples by denaturing gradient gel electrophoresis of 16S rDNA fragments. *Arch. Microbiol.* **164:**165–172.

Myers, R. M., Maniatis, T., and Lerman, L. S. 1987. Detection and localization of single base changes by denaturing gradient gel electrophoresis. *Methods Enzymol.* **155:**501–527.

Nold, S. C., Kopczynski, E. P., and Ward, D. M. 1996. Cultivation of proteobacteria and gram positive bacteria from a hot spring microbial mat. *Appl. Environ. Microbiol.* **62:**3917–3921.

Nold, S. C., and Ward, D. M. 1995. Diverse *Thermus* species inhabit a single hot spring microbial mat. *Syst. Appl. Microbiol.* **18:**274–278.

Nübel, U., Engelen, B., Felske, A., Snaidr, J., Wreshuber, A., Amann, R. I., and Ludwig, W. 1996. Sequence heterogeneities of genes encoding 16S rRNAs in *Paenibacillus polymyxa* detected by temperature gradient gel electrophoresis. *J. Bacteriol.* **178:**5636–5643.

Olsen, G. J., Lane, D. J., Giovannoni, S. J., and Pace, N. R. 1986. Microbial ecology and evolution: A ribosomal RNA approach. *Annu. Rev. Microbiol.* **40:**337–365.

Olsen, G. J., and Woese, C. R. 1993. Ribosomal RNA: A key to phylogeny. *FASEB J.* **7:**113–123.

Pace, N. R., Stahl, D. A., Lane, D. J., and Olsen, G. J. 1986. The analysis of natural microbial populations by ribosomal RNA sequences. *Adv. Microb. Ecol.* **9:**1–55.

Pierson, B. K., and Castenholz, R. W. 1974. A phototrophic gliding filamentous bacterium of hot springs, *Chloroflexus aurantiacus*, gen. and sp. nov. *Arch. Microbiol.* **100:**5–24.

Rainey, F. A., Ward, N., Sly, L. I., and Stackebrandt, E. 1994. Dependence of the taxonomic composition of clone libraries from PCR-amplified naturally occurring 16S rDNA on the primer pair and the cloning system used. *Experientia* **50:**789–801.

Rainey, F. A., Ward-Rainey, N. L., and Stackebrandt, E. 1996. *Clostridium paradoxum* DSM 7308T contains multiple 16S rRNA genes with heterogeneous intervening sequences. *Microbiology* **142:**2087–2091.

Ramsing, N. B., Ferris, M. J., and Ward, D. M. 2000. Highly ordered vertical structure of *Synechococcus* populations within the one-millimeter-thick photic zone of a hot spring cyanobacterial mat. *Appl. Environ. Microbiol.* **66:**1038–1049.

Reysenbach, A. L., Giver, L. J., Wickham, G. S., and Pace, N. R. 1992. Differential amplification of rRNA genes by polymerase chain reaction. *Appl. Environ. Microbiol.* **58:**3417–3418.

Robison-Cox, J. F., Bateson, M. M., and Ward, D. M. 1995. Evaluation of nearest-neighbor methods for detection of chimeric small-subunit rRNA sequences. *Appl. Environ. Microbiol.* **61:**1240–1245.

Ruff-Roberts, A. L., Kuenen, G. J., and Ward, D. M. 1994. Distribution of cultivated and uncultivated cyanobacteria and *Chloroflexus*-like bacteria in hot spring microbial mats. *Appl. Environ. Microbiol.* **60:**697–704.

Sandbeck, K. A., and Ward, D. M. 1982. Temperature adaptations in the terminal processes of anaerobic decomposition of Yellowstone and Icelandic hot spring microbial mats. *Appl. Environ. Microbiol.* **44:**844–851.

Santegoeds, C. M., Nold, S. C., and Ward, D. M. 1996. Denaturing gradient gel electrophoresis used to monitor the enrichment culture of aerobic chemoorganotrophic bacteria from a hot spring cyanobacterial mat. *Appl. Environ. Microbiol.* **58:**3417–3418.

Schink, B., and Zeikus, J. G. 1983. *Clostridium thermosulfurogenes* sp. nov., a new thermophile that produces elemental sulphur from thiosulphate. *J. Gen. Microbiol.* **129:**1149–1158.

Schmidt, T. M., DeLong, E. F., and Pace, N. R. 1991. Analysis of marine picoplankton community by 16S rRNA gene cloning and sequencing. *J. Bacteriol.* **173:**4371–4378.

Schulze, E., and Mooney, H. A. 1993. *Biodiversity and ecosystem function.* Heidelberg: Springer-Verlag.

Stackebrandt, E., Liesack, W., and Goebel, B. M. 1993. Bacterial diversity in a soil sample from a subtropical Australian environment as determined by 16S rDNA analysis. *FASEB J.* **7:**232–236.

Suzuki, M. T., Rappe, M. S., Haimberger, Z. W., Winfield, H., Adair, N., Strobel, J., and Giovannoni, S. J. 1997. Bacterial diversity among small-subunit rRNA gene clones and cellular isolates from the same seawater sample. *Appl. Environ. Microbiol.* **63:**983–989.

Ward, D. M., Bateson, M. M., Weller, R., and Ruff-Roberts, A. L. 1992. Ribosomal RNA analysis of microorganisms as they occur in nature. In Marshal, K. C. (ed.) *Advances in microbial ecology* (pp. 219–286). New York: Plenum.

Ward, D. M., Ferris, M. J., Nold, S. C., Bateson, M. M., Kopczynski, E. D., and Ruff-Roberts, A. L. 1994. Species diversity in hot spring microbial mats as revealed by both molecular and enrichment culture approaches-relationship between biodiversity and community structure. In Stal, L. J., and Caumette, P. (eds.) *Microbial mats: Structure, development and environmental significance* (pp. 33–44). Berlin: Springer-Verlag.

Ward, D. M., Santegoeds, C. M., Nold, S. C., Ramsing, N. B., Ferris, M. J., and Bateson, M. M. 1997. Biodiversity within hot spring microbial mat communities: Molecular monitoring of enrichment cultures. *Antonie Leeuwenhoek* **71:**143–150.

Ward, D. M., Tayne, T. A., Anderson, K. L., and Bateson, M. M. 1987. Community structure and interactions among community members in hot spring cyanobacterial mats. *Symp. Soc. Gen. Microbiol.* **41:**179–210.

Ward, D. M., Weller, R., and Bateson, M. M. 1990a. 16S rRNA sequences reveal numerous uncultured microorganisms in a natural community. *Nature* **345:**63–65.

Ward, D. M., Weller, R., and Bateson, M. M. 1990b. 16S rRNA sequences reveal uncultured inhabitants of a well-studied thermal community. *FEMS Microbiol. Rev.* **75:**105–116.

Ward, D. M., Weller, R., Shiea, J., Castenholz, R. W., and Cohen, Y. 1989. Hot spring microbial mats: Anoxygenic and oxygenic mats of possible evolutionary significance. In Cohen, Y., and Rosenberg, E. (eds.) *Microbial mats physiological ecology of benthic microbial communities.* Washington, D.C.: American Society for Microbiology.

Wawer, C., and Muyzer, G. 1995. Genetic diversity of *Desulfovibrio* spp. in environmental samples analyzed by denaturing gradient gel electrophoresis of [NiFe] hydrogenase gene fragments. *Appl. Environ. Microbiol.* **61:**2203–2210.

Weller, R., Bateson, M. M., Heimbuch, B. K., Kopczynski, E. D., and Ward, D. M. 1992. Uncultivated cyanobacteria, *Chloroflexus*-like inhabitants, and spirochete-like inhabitants of a hot spring microbial mat. *Appl. Environ. Microbiol.* **58:**3964–3969.

Weller, R., Weller, J. W., and Ward, D. M. 1991. 16S rRNA sequences of uncultivated hot spring cyanobacterial mat inhabitants retrieved as randomly primed cDNA. *Appl. Environ. Microbiol.* **57:**1146–1151.

Widdel, F., Schnell, S., Heising, S., Ehrenreich, A., Assmus, B., and Schink, B. 1993. Ferrous iron oxidation by anoxygenic phototrophic bacteria. *Nature* **362:**834–836.

Wiegel, J., Braun, M., and Gottschalk, G. 1981. *Clostridium thermoautotrophicum* species novum, a thermophile producing acetate from molecular hydrogen and carbon dioxide. *Curr. Microbiol.* **5:**255–260.

Wiegel, J., and Ljungdahl, L. G. 1981. *Thermoanaerobium ethanolicus* gen. nov., sp. nov., a new, extreme thermophilic, anaerobic bacterium. *Arch. Microbiol.* **128:**343–348.

Wiegel, J., Ljungdahl, L. G., and Rawson, J. R. 1979. Isolation from soil and properties of the extreme thermophile *Clostridium thermohydrosulfuricum. J. Bacteriol.* **139:**800–810.

Woese, C. R. 1987. Bacterial evolution. *Microbiol. Rev.* **51:**221–271.

Woese, C. R. 1994. Microbiology in transition. *Proc. Natl. Acad. Sci. USA* **91:**1601–1603.

Zeikus, J. G., Ben-Basset, A., and Hegge, P. W. 1980. Microbiology of methanogenesis in thermal, volcanic environments. *J. Bacteriol.* **143:**432–440.

Zeikus, J. G., Dawson, M. A., Thompson, T. E., Ingvorsen, K., and Hatchikian, E. C. 1983. Microbial ecology of volcanic sulphidogenesis: Isolation and characterization of *Thermodesulfobacterium commune* gen. nov. and sp. nov. *Arch. Microbiol.* **129:**1159–1169.

Zeikus, J. G., Gegge, P. W., and Anderson, M. A. 1979. *Thermoanaerobium brockii* gen. nov. and sp. nov., a new chemoorganotrophic, caldoactive, anaerobic bacterium. *Arch. Microbiol.* **122:**41–48.

Zeikus, J. G., and Wolfe, R. S. 1972. *Methanobacterium thermoautotrophicum* sp. nov., an anaerobic, autotrophic, extreme thermophile. *J. Bacteriol.* **109:**707–713.

Zeng, Y. B., Ward, D. M., Brassell, S., and Eglinton, G. 1992a. Biogeochemistry of hot spring environments. 3. Apolar and polar lipids in the biologically active layers of a cyanobacterial mat. *Chem. Geol.* **95:**347–360.

Zeng, Y. B., Ward, D. M., Brassell, S., and Eglinton, G. 1992b. Biogeochemistry of hot spring environments. 2. Lipid compositions of Yellowstone (Wyoming, USA) cyanobacterial and *Chloroflexus* mats. *Chem. Geol.* **95:**327–345.

6

Direct 5S rRNA Assay for Microbial Community Characterization

D. L. Stoner, C. K. Browning, D. K. Bulmer, T. E. Ward, and M. T. MacDonell

1. INTRODUCTION

Research in the geothermal regions of Yellowstone National Park has revealed an abundance of microorganisms that inhabit these extreme environments. The description of these microbial populations has been addressed by a number of approaches. One approach was the cultivation and identification of microorganisms using classical microbiological methods (Brock and Darland, 1971; Brock and Freeze, 1969; Brock et al., 1972). Other approaches use molecular methods that identify microorganisms by sequence analysis of nucleic acids extracted from collected biomass (Stahl et al., 1985), polymerase chain reaction (PCR)-amplified rRNA genes from clones and isolates (Barns et al., 1994; Reysenbach et al., 1994; Saul et al., 1993), and cDNA synthesized from small subunit ribosomal RNA (16S rRNA) extracted from collected biomass (Ward et al., 1990).

A variety of techniques has been used to assess the ecological role of microorganisms in geothermal environments. The overall goal was to elucidate metabolic pathways and determine the flux of carbon and energy through the system. The classical approach is the study of microorganisms in culture (Zeikus et al., 1980). By correlating the information obtained from pure culture studies with field observations, potential ecological functions can be proposed. Population-based physiological studies such as those that assess primary

D. L. Stoner, D. K. Bulmer, and T. E. Ward • Biotechnology, Idaho National Engineering and Environmental Laboratory, Idaho Falls, Idaho 83415-2203. **C. K. Browning and M. T. MacDonell** • Ransom Hill Bioscience, Inc., Ramona, California 92065.

Thermophiles: Biodiversity, Ecology, and Evolution, edited by Reysenbach *et al.* Kluwer Academic/Plenum Publishers, New York, 2001.

productivity (Doemel and Brock, 1977; Revsbech and Ward, 1984) or the transformation of organic materials (Anderson et al., 1987; Stoner et al., 1994; Ward, 1978) assess the overall capability of a microbial consortium. The role of broad physiological classes of microorganisms, for example, sulfate-reducing bacteria, methanogens, photoautotrophs, can be deduced by changes in microbial activity caused by the addition of metabolic inhibitors (Anderson et al., 1987; Stoner et al., 1994) or the exclusion of light (Revsbech and Ward, 1984). Molecular approaches in conjunction with physiological studies have been used to assign ecological roles to microorganisms in Yellowstone springs (Bateson and Ward, 1995; Ruff-Roberts et al., 1994). Oligonucleotide hybridization probes which are complementary to small subunit rRNA sequences were used to study the changes in microbial mat populations after they were shifted, *in situ*, to a region of the spring system that had a different temperature (Ruff-Roberts et al., 1994). Cultivation techniques in conjunction with 16S rDNA molecular analysis has been used to analyze the dominant members of mat communities (Bateson and Ward, 1995).

Molecular and cultivation techniques may make it difficult to determine whether a particular species is an active participant in a microbial community, is simply passive, or is transitory. A passive state, called senescence, has been well characterized in bacteria isolated from certain natural environments (Roszak et al., 1984). The precise role of senescent or dormant microorganisms in microbial communities is unknown, but it has been speculated that they may comprise a natural reservoir that can react rapidly to the onset of conditions most favorable to them. The impact of "senescent" strains on attempts to characterize environmental microbial populations often exhibits itself as a marked disparity between plate counts and direct counts (Byrd et al., 1991; Ghiorse and Balkwill, 1983).

An understanding of the link between community structure and function necessitates assessing the active microbial population. Until a comparative study is undertaken, the relative merits of a 23S-, 16S-, and 5S rRNA-based assessment of *in situ* microbial activity is only open to speculation. Although the small size of the 5S rRNA molecule is sufficient for many ecological studies, it imposes some constraint upon the phylogenetic information that is available. However, extensive detailed phylogenetic information is not necessary for a wide variety of ecological studies. The information that is provided by 5S rRNA is sufficient to answer many questions, including which members of the microbial community are active based on their ribosomal RNA abundance. In many cases, the interpretation of ecological function would not differ significantly when a community member and its associated metabolic activity is discussed in terms of genus, for example, *Leptospirillum*; species, for example, *Leptospirillum ferrooxidans*; or strain, for example, *Leptospirillum ferrooxidans* MK. Furthermore, even though the most minor component of a microbial population can be identified by PCR-amplification techniques, it is very difficult to analytically determine the geochemical transformation rates associated with that minor community member.

We investigated the use of denaturing gradient gel electrophoresis (DGGE) of 5S rRNA extracted from biomass to assess the predominant, active members of a microbial community. This study reports data that were acquired from the 5S rRNA DGGE analysis of field samples collected from acidic springs within the Norris Geyser region of Yellowstone National Park. The advantage of the 5S rRNA DGGE approach is that it is simple and rapid and, when fully developed, could be used to identify the dominant, active members of a microbial population.

2. MATERIALS AND METHODS

2.1. Microorganisms

The bacterial cultures used are listed in Table 1. The *Acidiphilium* spp. and the moderately thermophilic bacteria designated YTH13 and YTH15 were cultivated in a modification of the glycerol salts medium of Johnson et al. (Johnson et al., 1987). The medium (pH 3) contained 1.25 g $(NH_4)_2SO_4$, 0.50 g $MgSO_4 \cdot 7H_2O$, 0.25 g trypticase soy broth (Difco, Detroit, MI), 1.8 g glycerol, and 0.5 mL trace elements solution, in 1 liter deionized water. The iron-oxidizing heterotroph designated SCL2 was cultivated in glycerol salts medium containing 50 mM $FeSO_4$. *Thiobacillus thiooxidans* was cultivated in a medium (Tuovinen and Kelly, 1974) at pH 3.5 that contained 3.0 g KH_2PO_4, 3.0 g $(NH_4)_2SO_4$, 0.50 g $MgSO_4 \cdot 7H_2O$, 0.25 g $CaCl_2 \cdot 2H_2O$, and 3.2 g $K_2S_4O_6$, in 1 L deionized water. *Thiobacillus ferrooxidans* and *Leptospirillum ferrooxidans* were cultivated in a basal salts medium at pH 2 (Johnson et al., 1987) that contained 1.25 g $(NH_4)_2SO_4$, 0.50 g $MgSO_4 \cdot 7H_2O$, 0.25 g K_2HPO_4, and 0.5 mL trace elements solution, in 1 L deionized water, using 50 mM $FeSO_4$ as the energy source. Mesophilic cultures were incubated at 30°C, and the moderate thermophiles at 45°C.

2.2. RNA Extraction

RNA was extracted by the method of Peppel and Baglioni (1990). For the *Acidiphilium* spp. and the moderate thermophiles, biomass was collected by centrifugation and approximately 0.1 mL or 2 loopfuls of material was transferred to 0.5 mL lysis solution (2% sodium dodecyl sulfate, 200 mM Tris-HCl, pH 7.5, 1 mM EDTA) and incubated at room

Table 1
Acidophilic Autotrophic and Heterotrophic Microorganisms

Culture	Original source	Reference
Acidiphilium angustum (KLB)	Mine drainage, Pennsylvania, U.S.A.	Wichlacz et al., 1986
Acidiphilium cryptum (Lhet2)	Culture of *Thiobacillus ferrooxidans*	Harrison, 1980
Acidiphilium facilis ATCC 35904 (PW2)	Abandoned mine, Pennsylvania, U.S.A.	Wichlacz and Unz, 1981
Leptospirillum ferroxidans (BU-1)	Copper mine drainage, Bulgaria	Harrison, 1986
Thiobacillus ferrooxidans ATCC 19859	Copper leachate, British Columbia, Canada	Lazaroff, 1963
Thiobacillus thiooxidans ATCC 8085	Soil	Vischniac and Santer, 1957
Thermophile heterotroph (YTH13)	Yellowstone National Park, U.S.A.	Johnson et al., 2001
Thermophile heterotroph (YTH15)	Yellowstone National Park, U.S.A.	Johnson et al., 2001
Iron-oxidizing heterotroph (SLC2)	Regolith sample, U.S.A.	Johnson, 1995

temperature for 5–10 min. After lysis, 150 μL of 5 M potassium acetate solution was added to each microcentrifuge tube, and the contenst were mixed. Following chilling on ice at −20° for 5–10 min, the solutions were clarified by centrifugation at maximum speed for 5 min at 5°C. Supernatants were removed, placed in clean microcentrifuge tubes, and extracted with 0.3 mL chloroform:isoamyl alcohol (24:1). The aqueous phase was transferred to a fresh tube, and nucleic acids were precipitated by adding 0.6 mL ice-cold isopropanol. The material was held at −20°C for a minimum of 20 min. RNA was suspended in diethylpyrocarbonate-treated deionized water, and the amount of nucleic acid was estimated by using DNA Quik STRIPs (Kodak, Laboratory Research Products, New Haven, CT).

The biomass from 5- to 10-L cultures *T. ferrooxidans*, *T. thiooxidans*, *L. ferrooxidans*, and the iron-oxidizing heterotroph, SLC2, was collected by filtration using 0.2-μm HT Tuffryn filter cartridges (Gelman Sciences, Ann Arbor, MI). For Yellowstone samples, biomass was collected on-site by filtering spring water onto HT Tuffryn filter cartridges, cooling on ice, then transporting back to the laboratory for processing within 4–6 hours. Cells were lysed on the filters with 10–15 mls of lysis solution; filter units were rolled intermittently to ensure thorough mixing. Lysed material was withdrawn from the inlet port of the filter unit and split between two 50-mL centrifuge tubes. The rest of the RNA preparation procedure continued as described above but with proportionally higher volumes of reagents, and the nucleic acid precipitation step used isopropanol overnight at −20°C.

2.3. Denaturing Gradient Gel Electrophoresis

A modification of the DGGE method of Myers et al. (1985, 1987) was used. Gels (14.5 × 20cm) whose urea and acrylamide gradients were perpendicular to the direction of electrophoretic separation were formed by casting gels then rotating them 90° for running. A gradient between 7M urea, 10% acrylamide (37.5:1) in tris-acetate-EDTA (TAE) buffer (40 mM Tris, 1 mM EDTA, and 20 mM acetate) and 15% acrylamide (37.5:1) in TAE was formed over half the gel, and the remaining half of the gel was formed from 15% acrylamide (37.5:1) in TAE. RNA preparations (15–50 μg nucleic acid) were distributed within the single well that ran along the entire top edge of the gel. Electrophoresis was performed in a water bath at a temperature of 65°C, for 5 h at 250V, constant voltage, using a Model-2000 dual capacity DGGE system (C.B.S. Scientific Company, Inc., Del Mar, CA) and a Bio-Rad 3000xi power supply (Bio-Rad Laboratories, Inc., Hercules, CA). Gels were stained with SYBR Green II RNA gel stain (Molecular Probes, Eugene, OR) in TAE, visualized with a UV transilluminator (Model T1202, Sigma Chemical Co., St. Louis, MO), and photographed with a Polaroid MP-4 Land Camera (Model 44-01, Polaroid Corporation, Cambridge, MA) using a SYBR Green gel photographic filter (Molecular Probes, Inc., Eugene, OR).

3. RESULTS

The 5S rRNA assay is based on separating 5S rRNA species by denaturing gradient gel electrophoresis (Stoner et al., 1996). Polyacrylamide gels whose gradients of urea and acrylamide were perpendicular to the direction of migration were formed such that the urea concentration decreased and the acrylamide concentration increased from the right edge to the middle of the gel. The left half of the gel contained 15% acrylamide. The samples were distributed within a single well that ran the entire width of gel. This configuration permits separating the 5S rRNAs along a continuum of changing acrylamide and urea concentrations. The 5S rRNAs and tRNAs (Bidle and Fletcher, 1995; Höfle, 1988) were discernible within the lower region of the gels, and the larger rRNAs were visible in the uppermost portion of the gels.

The 5S rRNA assay was evaluated by using microorganisms whose origins were acidic mining environments (Stoner et al., 1996). Figures 1, 3, 4, and 5 have been discussed previously (Stoner et al., 1996) and are provided herein for comparison. Resolution, using DGGE analysis of 5S rRNA, was demonstrated to be effective to the species level (Stoner et al., 1996). The assay can resolve the 5S rRNA of *Thiobacillus ferrooxidans* (ATCC 19859) and *Thiobacillus thiooxidans* (ATCC 8085) (Fig. 1), whose sequences differ at only

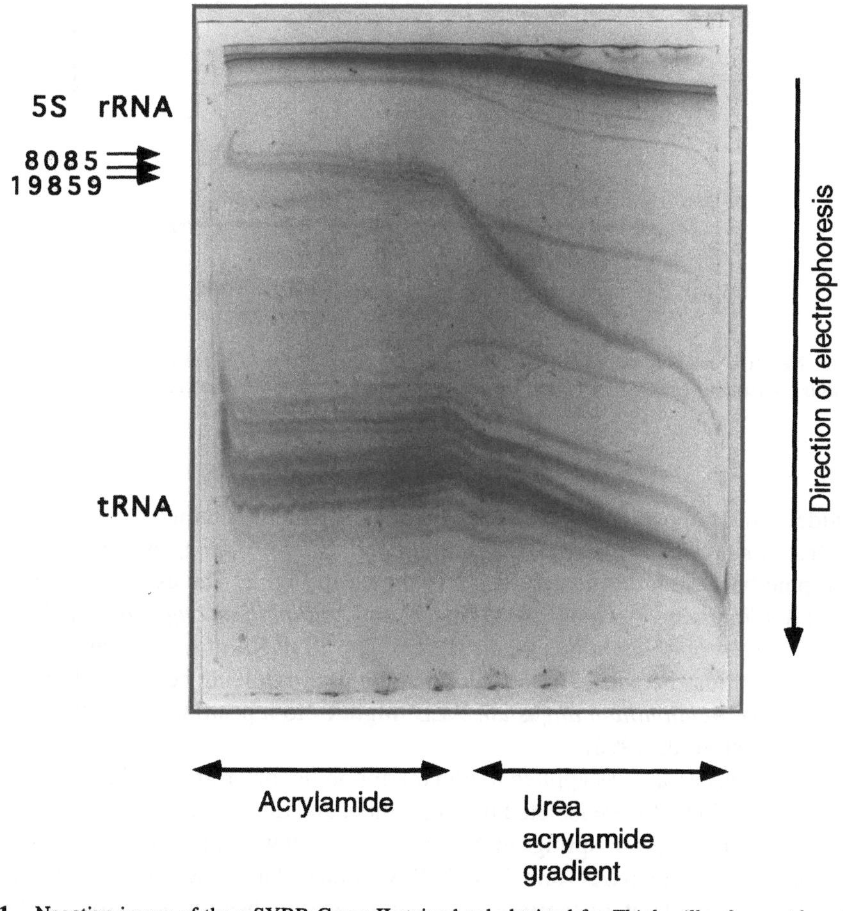

Figure 1. Negative image of the a SYBR Green II stained gel obtained for *Thiobacillus ferrooxidans* (ATCC 19859) and *Thiobacillus thiooxidans* (ATCC 8085).

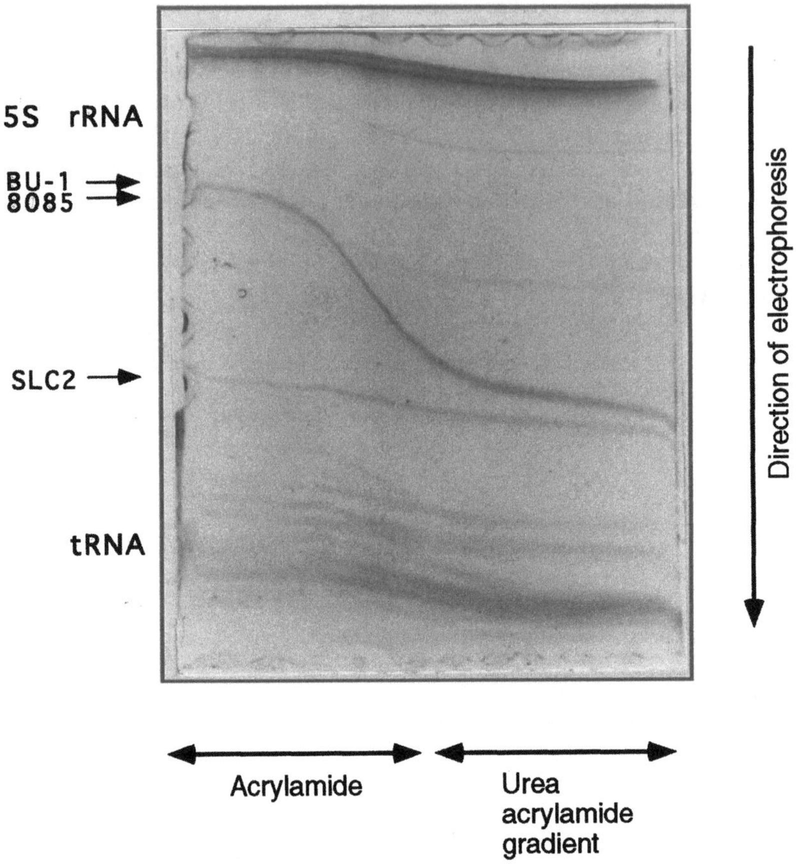

Figure 2. Negative image of a SYBR Green II stained gel obtained for *Leptospirillum ferrooxidans* (BU-1), *Thiobacillus thiooxidans* (ATCC 8085), and an iron-oxidizing heterotrophic bacterium designated SLC2.

three widely spaced positions (Lane et al., 1985). The pattern obtained for a mixture of 5S rRNAs from *Leptospirillum ferrooxidans, Thiobacillus thiooxidans*, and an iron-oxidizing heterotrophic bacterium designated SLC2 is shown in Fig. 2. The assay resolved the 5S rRNAs of *Acidiphilium cryptum* (Lhet2) (Fig. 3) and *Acidiphilium angustum* (KLB) (Fig. 4) from the 5S rRNA of *Acidiphilium facilis* (PW2). The 5S rRNA of *A. cryptum* and *A. facilis* migrated to positions between *Thiobacillus* and the iron-oxidizing heterotroph, SLC2, and the 5S rRNA of *Acidiphilium angustum* KLB migrates to a position above that of *Leptospirillum* (Stoner et al., 1996).

The 5S rRNA assay was applied to two thermophilic, heterotrophic bacterial cultures, YTH13 and YTH15 that were isolated by enrichment and direct plating, respectively, from water collected from Sylvan Springs in the Gibbons Meadows region (Johnson et al., 2001). The positions of the 5S rRNAs from these Yellowstone isolates (Fig. 5) were similar to that for SLC2 and a moderately thermophilic iron-oxidizing heterotroph isolated from an industrial heap leach operation (Stoner et al., 1996). The patterns of the 5S rRNAs for these

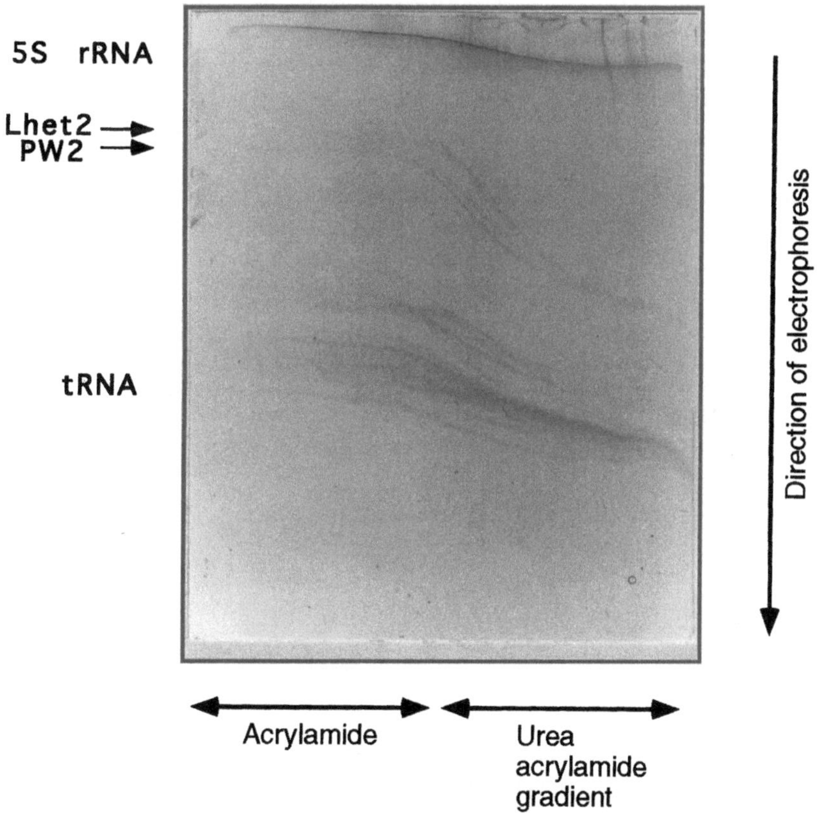

Figure 3. Negative image of a SYBR Green II stained gel obtained for *Acidiphilium cryptum* (Lhet2) and *Acidiphilium facilis* (PW2).

microorganisms were distinctly different from those of all of the gram-negative bacteria that were evaluated. The 16S rRNA sequences of these newly isolated microorganisms (Johnson et al., 2001) resemble those of gram-positive microorganisms; however, their gram-staining reactions were indeterminate (Roberto, personal communication). The comigration of the 5S rRNAs from YTH13 and YTH15 may suggest that these clones are closely related. This is supported by 16S rRNA sequence data analysis (Roberto, personal communication).

The 5S rRNA profile (Fig. 6) obtained from the source water of a small spring in the Hundred Springs region of the Norris Geyser Basin showed one dominant band and one minor band, whereas a sample retrieved down channel from this source exhibited three distinct bands (Fig. 7). Two other springs within the Norris Geyser Basin district also exhibited simple 5S rRNA profiles (Figs. 8 and 9). Bands suggesting *Leptospirillum* and *Thiobacillus* 5S rRNA were absent even though there was presumptive evidence of sulfur (Fig. 8) and iron (Fig. 9) precipitates in these springs. The bands that were detected for these

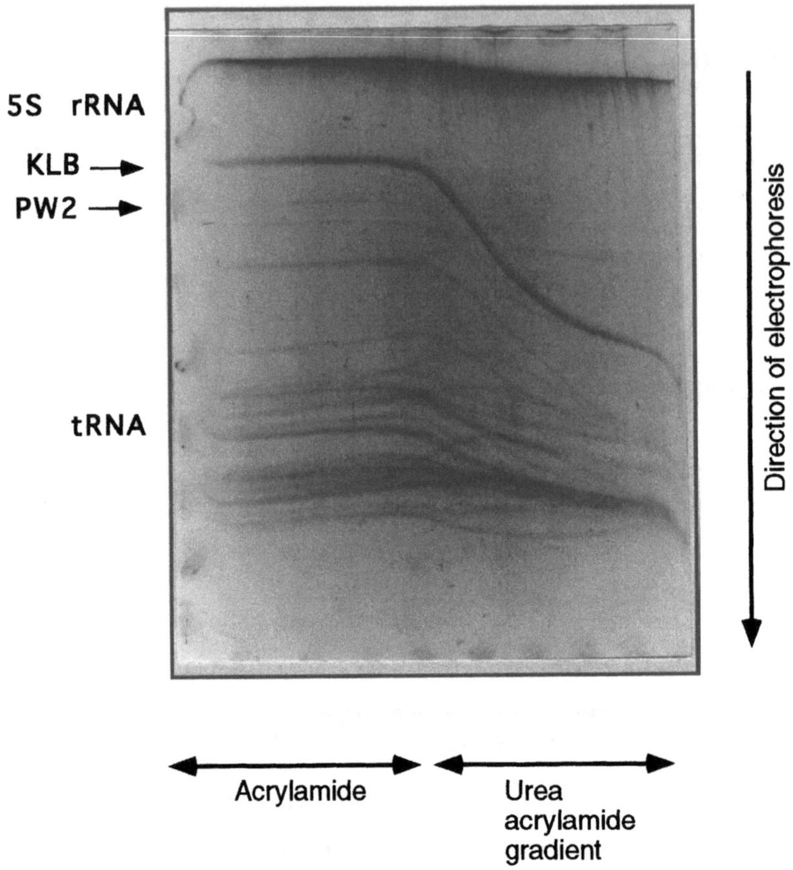

Figure 4. Negative image of a SYBR Green II stained gel obtained for *Acidiphilium angustum* (KLB) and *Acidiphilium facilis* (PW2).

spring samples migrated to the lower portion of the gel, which suggests a predominantly gram-positive bacterial community. The 5S rRNA of *Clostridium* sp., *Bacillus cereus*, *Bacillus subtilis*, and a "gram-positive-like" isolate from an acidic mining environment migrate to this lower region of the gel (data not shown). The resolution of the assay has been determined to be to the species level for gram-negative bacteria. However, whether the resolution of this assay extends to the regions of the gel to which 5S rRNAs of gram-positive or gram-positive-like microorganisms migrated still needs to be evaluated in further studies.

4. DISCUSSION

The DGGE analysis of 5S rRNA described herein is a relatively simple and rapid technique that can be used to assess the active members of a microbial population. For a

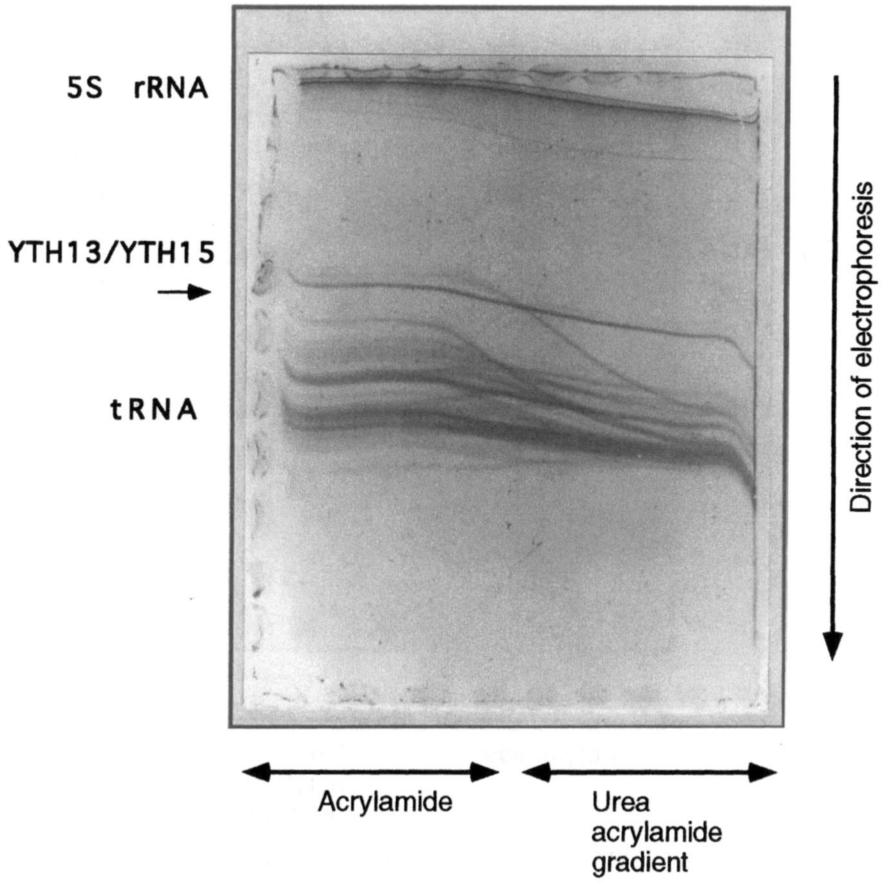

Figure 5. Negative image of a SYBR Green II stained gel obtained for the moderately thermophilic heterotrophic bacteria designated YTH13 and YTH15.

moderate level of microbial community detail, 5S rRNA analysis is superior to most other techniques, provided enough sample can be obtained. In this technique, macromolecules directly from the bacteria are analyzed, unlike the case for other techniques. 5S rRNA molecules are small enough so that those that have different sequences exhibit significant and useful differences in electrophoretic behavior, unlike whole 16 and 23S rRNA molecules, for which no suitable technique presently exists to separate those derived from closely related bacteria. In addition, the 5S rRNA analysis is not subject to the variety of amplification artifacts that are inherent in PCR amplification strategies, especially when analyzing substrate DNA that have targets different in abundance and sequence. Differential efficiency of amplifying the various sequences leads to complete loss of some sequences as well as significant alterations in the relative abundance of sequences in the amplified products compared to those in the original sample (Ferris and Ward, 1997), whose distribution one is trying to measure.

The results of this study suggested only a few species are the major components of the

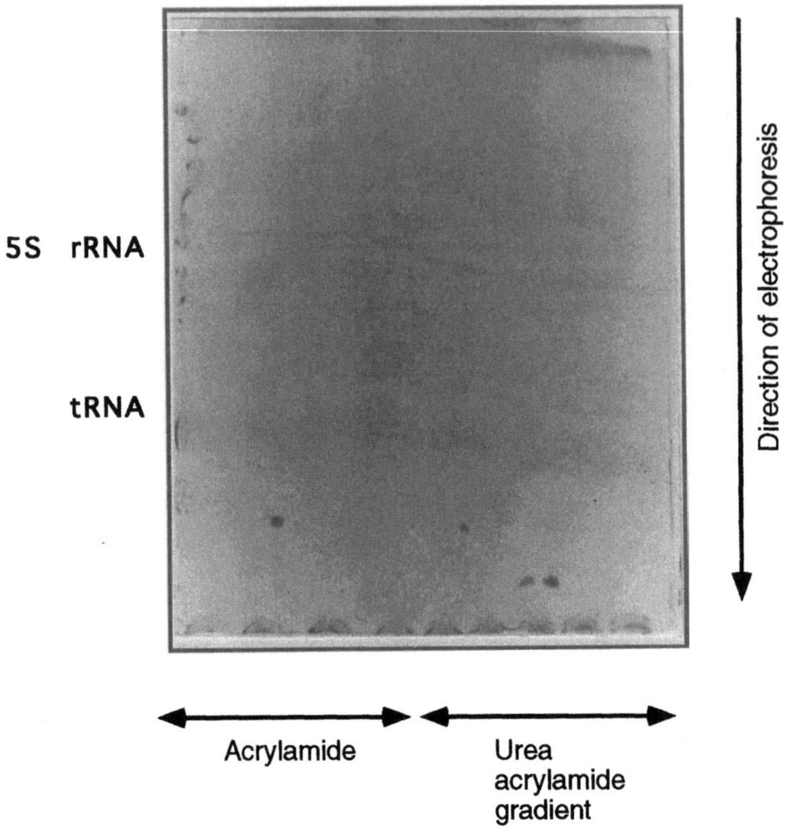

Figure 6. Negative image of a SYBR Green II stained gel obtained for ~10μg of nucleic acid extracted from biomass filtered from 12 L of water (65°C, pH 4). The small spring was located in the Hundred Springs region of Norris Geyser Basin.

microbial community within the waters of these geothermal springs. The lack of extensive diversity observed in various environmental samples by us and others (Ferris and Ward, 1997) may or may not be surprising, depending on one's preconceptions. It may reflect the phenomenon that, given relatively constant environmental conditions and the continual flushing within the spring ecosystems, the small number of bacteria that are best suited to multiply under those particular conditions eventually dominate the population of that environment. Many might suppose that there is an inherent limitation of the 5S rRNA technique described herein, but the amplification technique to detect 16S rRNA or other genes does not always result in more complex microbial profiles. Simple diversity profiles have been determined for a cyanobacterial community in an Octopus Spring effluent channel (Ferris and Ward, 1997) using DGGE analysis of PCR-amplified 16S rRNA gene fragments.

The different 5S rRNA profiles that were obtained for the source water of a small spring and water collected down channel suggested the presence of different populations

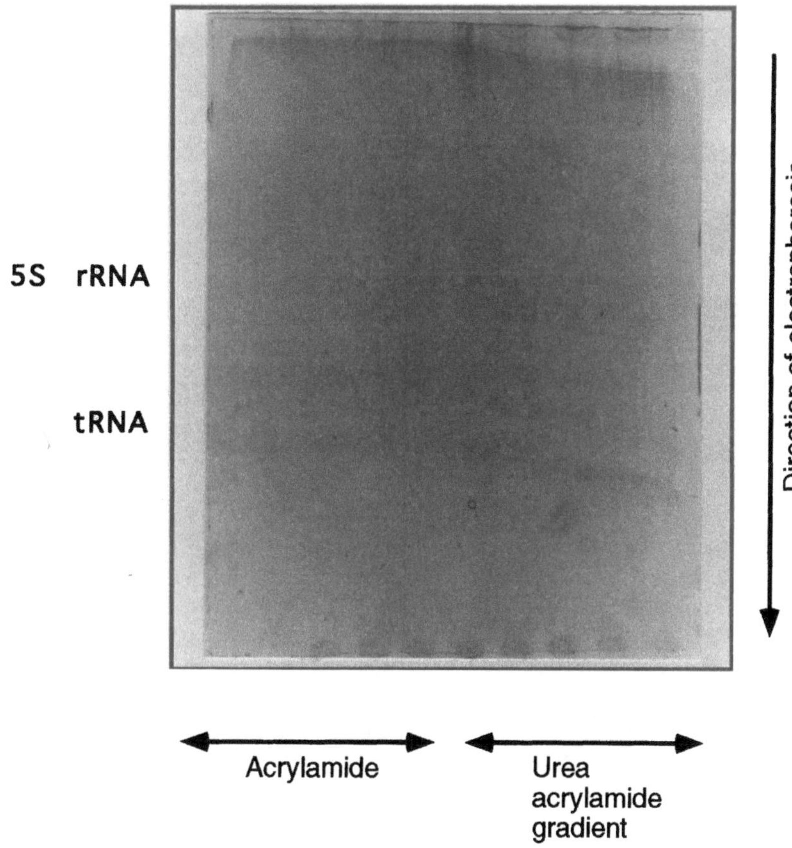

Figure 7. Negative image of a SYBR Green II stained gel obtained for 25 μg nucleic acid extracted from biomass filtered from ~10 L of water (38°C, pH 5) obtained 25–30 yards downstream from a small spring. The spring was located in the Hundred Springs region of Norris Geyser Basin.

that reflected the *in situ* conditions. Differences in microbial populations in relation to temperatures within Yellowstone hot spring environments has been documented (Brock and Brock, 1968; Ferris and Ward, 1997; Ruff-Roberts et al., 1994). In support of these studies, the 5S rRNA assay may provide an effective estimate of relative microbial activity and, thus, provide a more definitive link between temperature and microbial activity in these environments. A more comprehensive study might evaluate the diversity of a microbial community using several or more culture- and molecular-based techniques and correlate this information with the chemical and physical conditions and the biogeochemical transformations detected in that environment. Presently, many diversity studies only infer metabolic capabilities of the detected species from their relative position on the phylogenetic tree and link the detected microorganisms to their environment solely by pH and temperature.

For the Yellowstone samples, gel patterns were obtained the day after the biomass was collected. The minimum number of steps and resulting minimum sample manipulation

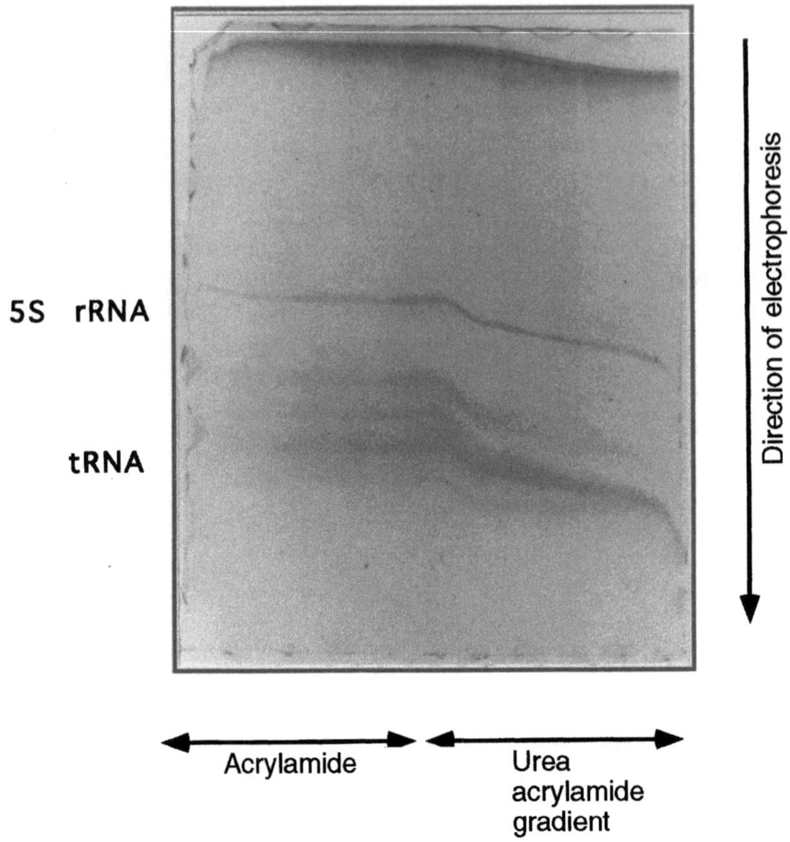

Figure 8. Negative image of a SYBR Green II stained gel obtained for ~50 μg of nucleic acid extracted from biomass filtered from 32 L of spring water (28°C, pH 3). The spring, bearing a sulfur-colored precipitate, was east of Hundred Springs in the Norris Geyser Basin.

before electrophoretic analysis reduces the potential for bias to enter the experiments. The only potential sources of bias in these experiments are the collection of biomass by filtration and the differential recovery of 5S rRNA from the various bacteria in the sample, which would be caused by different efficiencies of lysis of the various bacteria. Filtration can be a source of bias unless the filters retain even the smallest bacteria. Methods that require the recovery of nucleic acid can be biased toward the more easily lysed members of the community (Weller and Ward, 1989) unless the procedures used are effective for the most difficult to lyse microorganisms. Analysis of bulk rRNA extracted from collected biomass minimizes bias introduced by amplification procedures (Kopczynski et al., 1994; Liesack et al., 1991; Reysenbach et al., 1992), but it does increase the amount of biomass required for the analyses.

The complete characterization of an undefined microbial community would, of course, require identifying and enumerating member species. The 5S rRNA assay has the potential for providing the identity of community members using sequence analysis of

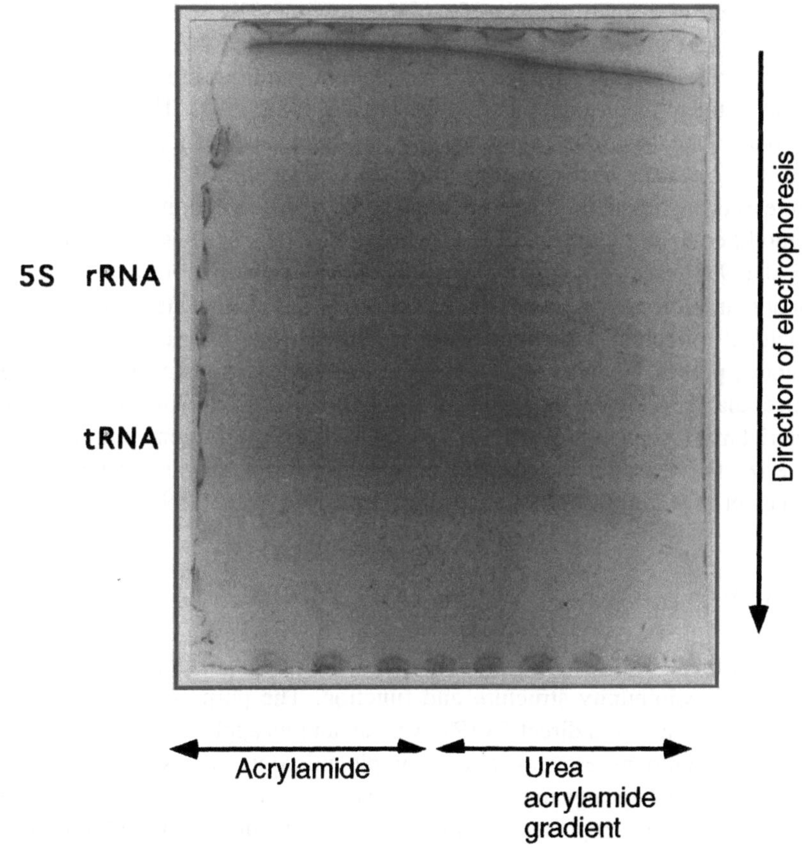

Figure 9. The negative image of a SYBR Green II stained gel obtained for ~15 μg nucleic acid extracted from biomass filtered from 32 L of water (42°C, pH 4.0). The spring that exhibited a rust-colored precipitate was located east of Sieve Lake in the Norris Geyser Basin.

individual 5S rRNA species that have been separated electrophoretically. This approach does not assume prior knowledge of community members, as do methods that rely upon using oligonucleotide probes, antibody-based detection, or nucleic acid hybridization techniques for identification. There are a number of examples in which rRNA species have been isolated directly from the environment, separated electrophoretically, and then sequenced for identification. Direct extraction and 5S rRNA sequence analysis was used to identify microorganisms from Octopus Spring, Yellowstone National Park, WY (Stahl et al., 1985); microorganisms in the mud obtained from a copper leaching pond (Lane et al., 1985); endosymbionts from a marine invertebrate (Pace et al., 1986); an antarctic endolithic bacterium (Colwell et al., 1989); and the dominant member of a freshwater microbial community (Höfle, 1992).

The 5S rRNA DGGE assay can provide an estimate of the relative abundance of community members by assessing the relative intensity of fluorescent dye stained 5S rRNA bands. The relative contribution of a particular 5S rRNA species to a microbial community could be the result of a small number of rapidly growing cells or a larger number of less active cells. Scans of autoradiograms of low molecular weight RNA (tRNA and 5S rRNA) profiles were used to monitor changes of microbial communities in freshwater mesocosms that were subjected to intentional perturbation (Höfle, 1988, 1992). This approach has also been used to compare the relative abundance of free-living and particle-associated bacterial communities in estuarine environments (Bidle and Fletcher, 1995).

If increased phylogenetic discrimination is desired, the conditions used for DGGE might be modified to allow the analysis of whole 16S rRNA species. In a related method, Muyzer et al. (1993) used DGGE to estimate genetic diversity in biofilms from polymerase chain-reaction amplified segments of the 16S rRNA genes. In that study, several members of an undefined microbial population were identified using hybridization analysis with oligonucleotide probes. Uncultivated Octopus Spring microorganisms have been identified by sequence analysis of cDNAs produced from 16S rRNA extracted from a hot spring cyanobacterial mat (Weller and Ward, 1989). DGGE of 16S rRNA gene fragments was used to characterize the seasonal distribution of species within a cyanobacterial mat in an effluent channel of a Yellowstone hot spring (Ferris and Ward, 1997).

5. SUMMARY

The assessment of microbial diversity and activity is integral to a fundamental understanding of community structure and function. The purpose of this study was to demonstrate the efficacy of a direct 5S rRNA assay and its application to the characterization of microbial communities using, as an example, microorganisms that inhabit the acidic effluents of geothermal springs in the Norris Geyser region of Yellowstone National Park. The direct 5S rRNA assay represents a direct, nonselective molecular method for monitoring and characterizing the metabolically active component of a microbial population. The assay involves using high-resolution denaturing gradient gel electrophoresis to separate 5S rRNA species extracted from collected biomass. Separation is based on the unique migratory behavior of each 5S rRNA species during electrophoresis in denaturing gradient gels. This study reports data that was acquired by DGGE analysis of 5S rRNA extracted from cultures of bacteria and field samples collected from acidic springs within Norris Geyser region of Yellowstone National Park. Discrimination, using this method, was effective to the species level. For example, the 5S rRNA from *Acidiphilium cryptum* was distinguished from that of *Acidiphilium facilis.* Two isolates obtained by enrichment from acidic Sylvan Springs, Gibbon Meadows region, migrated in a pattern that was distinct from the laboratory cultures tested. Banding patterns similar to those found for Sylvan Springs isolates were observed in acidic effluents of springs located in the Norris Geyser Basin. In conclusion, the 5S rRNA assay represents a powerful method for assessing active microbial populations. The four spring samples examined in this study, were the first four environmental samples that have been examined using this assay. The relative ease of success and the rapidity of obtaining data demonstrate the simplicity of this assay and its suitability for field studies. This study was not intended as an exhaustive examination of the microorganisms indigenous to acidic hot springs. Yet, even in this first application, the 5S rRNA assay suggested some areas of research that merit further study.

Acknowledgments. This work was supported by the Laboratory Directed Research and Development Program of the Idaho National Engineering and Environmental

Laboratory under contract number DE-AC01-94ID13223 with the Department of Energy, Idaho Operations Office. Funding was also provided by the Department of Energy, Basic Energy Sciences.

REFERENCES

Anderson, K. L., Tayne, T. A., and Ward, D. M. 1987. Formation and fate of fermentation products in hot spring cyanobacterial mats. *Appl. Environ. Microbiol.* **53**:2343.

Barns, S. M., Fundyga, R. E., Jeffries, M. W., and Pace, N. R. 1994. Remarkable archaeal diversity detected in Yellowstone National Park hot spring environment. *Proc. Natl. Acad. Sci. USA* **91**:1609.

Bateson, M. M., and Ward, D. M. 1995. Analysis of *Chloroflexus* diversity in hot spring microbial mats by molecular methods and extincting dilution enrichment. Abstracts of the *95th General Meeting of the American Society for Microbiology*, 1995. Washington, D.C.: American Society for Microbiology, p. 65.

Bidle, K. D., and Fletcher, M. 1995. Comparison of free-living and particle-associated bacterial communities in the Chesapeake Bay by stable low-molecular-weight RNA analysis. *Appl. Environ. Microbiol.* **61**:944.

Brock, T. D., and Brock, M. L. 1968. Relationship between environmental temperature and optimum temperature of bacteria along a hot spring thermal gradient. *J. Appl. Bacteriol.* **31**:54.

Brock, T. D., and Darland, G. K. 1971. *Bacillus acidocaldarius* sp. nov. an acidophilic thermophilic spore-forming bacterium. *J. Gen. Microbiol.* **67**:9.

Brock, T. D., and Freeze, H. 1969. *Thermus aquaticus* gen. nov. and sp. nov., a non-sporulating extreme thermophile. *J. Bacteriol.* **98**:289.

Brock, T. D., Brock, K. M., Belly, R. T., and Weiss, R. L. 1972. *Sulfolobus*: A new genus of sulfur-oxidizing bacteria living at low pH and high temperature. *Arch. Microbiol.* **84**:54.

Byrd, J. J., Xu, H. S., and Colwell, R. R. 1991. Viable but nonculturable bacteria in drinking water. *Appl. Environ. Microbiol.* **57**:875.

Colwell, R. R., MacDonell, M. T., and Swartz, D. G. 1989. Identification of an antarctic endolithic microorganism by 5S rRNA sequence analysis. *Syst. Appl. Microbiol.* **11**:182.

Doemel, W. N., and Brock, T. D. 1977. Structure, growth, and decomposition of laminated algal-bacterial mats in alkaline hot springs. *Appl. Environ. Microbiol.* **34**:433.

Ferris, M. J., and Ward, D. M. 1997. Seasonal distributions of dominant 16S rRNA-defined populations in a hot spring microbial mat examined by denaturing gradient gel electrophoresis. *Appl. Environ. Microbiol.* **63**: 1375–1381.

Ghiorse, W. C., and Balkwill, D. L. 1983. Enumeration and morphological characterization of bacteria indigenous to subsurface environments. *Dev. Ind. Microbiol.* **24**:213.

Harrison, A. P., Jarvis, B. W., and Johnson, J. L. 1980. Heterotrophic bacteria from cultures of autotrophic *Thiobacillus ferrooxidans*: Relationships as studied by means of deoxyribonucleic acid homology. *J. Bacteriol.* **143**:448.

Harrison, A. P. 1986. Characteristics of *Thiobacillus ferrooxidans* and other iron-oxidizing bacteria, with emphasis on nucleic acid analyses. *Biotechnol. Appl. Biochem.* **8**:249.

Höfle, M. G. 1988. Identification of bacteria by low molecular weight RNA profiles: A new chemotaxonomic approach. *J. Microbiol. Methods* **8**:235.

Höfle, M. G. 1992. Bacterioplankton community structure and dynamics after large-scale release of non-indigenous bacteria as revealed by low molecular weight RNA analysis. *Appl. Environ. Microbiol.* **58**:3387.

Johnson, D. B., personal communication.

Johnson, D. B., Bacelar-Nicolau, P., Bruhn, D. F., and Roberto, F. F. 1995. Iron-oxidizing heterotrophic acidophiles: Ubiquitous novel bacteria in leaching environments. In November 19–22, 1995, Vargas, T., Jerex, C. A., Wiertz, J. V., and Toledo, H. (eds.), *Biohydrometallurgical processing Vol. I. Proceedings of the International Biohydrometallurgy Symposium* IBS-95 (pp. 47–56). Viña del Mar, Chile: University of Chile.

Johnson, D. B., Body, D. A., Bridge, T. A. M., Bruhn, D. F., and Roberto, F. F. 2001. Biodiversity of acidophilic moderate thermophiles isolated from two sites in Yellowstone National Park, and their roles in the dissimilatory oxido-reduction of iron. In Reysenbach, A. L., Voytek, M., and Mancinelli, R. (eds.), *Thermophiles: Biodiversity, Ecology, and Evolution* (pp. 23–39). New York: Kluwer Academic/Plenum Press.

Johnson, D. B., Macvicar, J. H. M., and Rolfe, S. 1987. A new medium for the isolation and enumeration of *Thiobacillus ferrooxidans* and acidophilic heterotrophic bacteria. *J. Microbiol. Methods* **7**:9.

Kopczynski, E. D., Bateson, M. M., and Ward, D. M. 1994. Recognition of chimeric small-subunit ribosomal DNAs composed of genes from uncultivated microorganisms. *Appl. Env. Microbiol.* **60:**746.

Lane, D. J., Stahl, D. A., Olsen, G. J., Heller, D. J., and Pace, N. R. 1985. Phylogenetic analysis of the genera *Thiobacillus* and *Thiomicrospira* by 5S rRNA species. *J. Bacteriol.* **163:**75.

Liesack, W., Weland, H., and Stackebrandt, E. 1991. Potential risks of gene amplification by PCR as determined by 16S rDNA analysis of a mixed-culture of strict barophilic bacteria. *Microb. Ecol.* **21:**191.

Myers, R. M., Fischer, S. G., Lerman, L. S., and Maniatis, T. 1985. Nearly all single base substitutions in DNA fragments joined to a GC-clamp can be detected by denaturing gradient gel electrophoresis. *Nucleic Acid Res.* **13:**3131–3145.

Myers, R. M., Maniatis, T., and Lerman, L. S. 1987. Detection and localization of single base changes by denaturing gradient gel electrophoresis. *Methods Enzymol.* **155:**501–527.

Muyzer, G., de Waal, E. C., and Uitterlinden, A. G. 1993. Profiling of complex microbial populations by denaturing gradient gel electrophoresis analysis of polymerase chain reaction-amplified genes coding for 16S rRNA. *Appl. Environ. Microbiol.* **59:**695.

Pace, N. R., Stahl, D. A., Lane, D. J., and Olsen, G. J. 1986. The analysis of natural microbial populations by ribosomal RNA sequences. *Adv. Microb. Ecol.* **9:**1.

Peppel, K., and Baglioni, C. 1990. A simple and fast method to extract RNA from tissue culture cells. *BioTechniques* **9:**711.

Revsbech, N. P., and Ward, D. M. 1984. Microelectrode studies of interstitial water chemistry and photosynthetic activity in a hot spring microbial mat. *Appl. Environ. Microbiol.* **48:**270.

Reysenbach, A.-L., Giver, L. J., Wickham, G. S., and Pace, N. R. 1992. Differential amplification of rRNA genes by polymerase chain reaction. *Appl. Environ. Microbiol.* **58:**3417.

Reysenbach, A.-L., Wickham, G. S., and Pace, N. R. 1994. Phylogenetic analysis of the hyperthermophilic pink filament community in Octopus Spring. Yellowstone National Park. *Appl. Environ. Microbiol.* **60:**2113.

Roberto, F. F., personal communication.

Roszak, D. B., Grimes, D. J., and Colwell, R. R. 1984. Viable but non-recoverable stage of *Salmonella enteriditis* in aquatic systems. *Can. J. Microbiol.* **30:**334.

Ruff-Roberts, A. L., Kuenen, J. G., and Ward, D. M. 1994. Distribution of cultivated and uncultivated cyanobacteria and *Chloroflexus*-like bacteria in hot spring microbial mats. *Appl. Environ. Microbiol.* **60:**697.

Saul, D. J., Rodrigo, A. G., Reeves, R. A., Williams, L. C., Borges, K. M., Morgan, H. W., and Bergquist, P. L. 1993. Phylogeny of twenty *Thermus* isolates constructed from 16S rRNA gene sequence data. *Int. J. Syst. Bacteriol.* **43:**754.

Stahl, D. A., Lane, D. J., Olsen, G. J., and Pace, N. R. 1985. Characterization of a Yellowstone hot spring microbial community by 5S rRNA sequences. *Appl. Environ. Microbiol.* **49:**1379.

Stoner, D. L., Burbank, N. S., and Miller, K. S. 1994. Anaerobic transformation of organosulfur compounds in microbial mats from Octopus Spring. *Geomicrobiol. J.* **12:**195.

Stoner, D. L., Browning, C. K., Bulmer, D. K., Ward, T. E., and MacDonell, M. T. 1996. Direct 5S rRNA for the characterization of mixed culture bioprocesses. *Appl. Environ. Microbiol.* **62:**1969.

Tuovinen, O. H., and Kelly, D. P. 1974. Studies on the growth of *Thiobacillus ferrooxidans* V. Factors affecting growth in liquid culture and development of colonies on solid media containing inorganic sulphur compounds. *Arch. Microbiol.* **98:**351.

Vischniac, W., and Santer, M. 1957. The thiobacilli. *Bacteriol. Rev.* **21:**195.

Ward, D. M. 1978. Thermophilic methanogenesis in a hot-spring algal-bacterial mat (71 to 30°C). *Appl. Environ. Microbiol.* **35:**1019.

Ward, D. M., Weller, R., and Bateson, M. M. 1990. 16S rRNA sequences reveal numerous uncultured microorganisms in a natural community. *Nature* **345:**63.

Weller, R., and Ward, D. M. 1989. Selective recovery of 16S rRNA sequences from natural microbial communities in the form of cDNA. *Appl. Environ. Microbiol.* **55:**1818.

Wichlacz, P. L., and Unz, R. F. 1981. Acidophilic, heterotrophic bacteria of acidic mine waters. *Appl. Environ. Microbiol.* **41:**1254.

Wichlacz, P. L., Unz, R. F., and Langworthy, T. A. 1986. *Acidiphilium angustum* sp. nov., *Acidiphilium facilis*, sp. nov., and *Acidiphilium rubrum*, sp. nov.: Acidophilic heterotrophic bacteria from acidic coal mine drainage. *Int. J. Syst. Bacteriol.* **36:**197.

Zeikus, J. G., Ben-Bassat, A., and Hegge, P. W. 1980. Microbiology of methanogenesis in thermal, volcanic environments. *J. Bacteriol.* **143:**432.

7

Community Structure along a Thermal Gradient in a Stream Near Obsidian Pool, Yellowstone National Park

Joseph R. Graber, Julie Kirshtein, Mark Speck, and Anna-Louise Reysenbach

1. INTRODUCTION

The study of the diversity of life on earth has traditionally been one of the cornerstones of biological science. Examination of variations among plant and animal species became the foundation for Charles Darwin's theory of evolution by natural selection. More recently, ecological studies have demonstrated that community diversity is an important determinant in ecosystem processes and stability (Tilman et al., 1994; Tilman et al., 1997). The study of the evolution and diversity of microorganisms, however, has lagged behind that of other life forms, largely because of the difficulties inherent in assessing microbial diversity in nature. Microorganisms lack the complex morphologies used to classify higher organisms, and shared physiological traits have also proven unreliable in inferring evolutionary relationships (Fox et al., 1980). Traditional culture techniques often lead to underestimation of diversity by selecting for those organisms best suited to the growth media (Brock, 1987).

In recent years, however, advances in molecular phylogenetic techniques have led to new approaches for examining environmental microbial diversity that circumvent some of

Joseph R. Graber, Julie Kirshtein, Mark Speck, and Anna-Louise Reysenbach • Department of Microbiology and Biochemistry, Cook College, Rutgers University, New Brunswick, New Jersey 08903.

Thermophiles: Biodiversity, Ecology, and Evolution, edited by Reysenbach *et al.* Kluwer Academic / Plenum Publishers, New York, 2001.

the problems associated with enrichment culture approaches. These techniques are based on the analysis of the small subunit rRNA (16S rRNA) gene sequences isolated directly from environmental samples (for reviews, see Amman et al., 1995 and Ward et al., 1992). By analyzing differences in 16S rRNA gene sequences, it is possible to differentiate among microbial species and infer evolutionary relationships among them (Woese, 1987). Oligonucleotide hybridization probes based on 16S rRNA sequence can then be used to trace organisms of interest *in situ* (DeLong et al., 1989). Examination of molecular phylogenies of life has had far reaching impact, mostly notably on the fundamental rethinking of the universal phylogenetic tree to reflect three domains, rather than five kingdoms, of life (Woese et al., 1990).

The diversity of thermophilic microorganisms that inhabit high temperature ecosystems, such as terrestrial hot springs and marine hydrothermal vents, has been a subject of increased scientific and biotechnological interest (Stetter et al., 1990). Thermophiles can grow at temperatures as high as 113°C (Blochl et al., 1997), and they possess a wide variety of metabolisms, including heterotrophy, phototrophy, and chemolithotrophy (Schonheit et al., 1995). These organisms are sources of interest for the biotechnology industry because various thermostable enzymes have been exploited for industrial applications (Adams et al., 1995). Furthermore, it has been proposed that the universal ancestor of all life on earth might have originated in a high temperature environment (Corliss, 1990; Woese et al., 1990). Thus, examining microbial life in extant high temperature environments may provide valuable insights into the early evolution and diversification of life on earth.

The prominent geothermal features of Yellowstone National Park (hot springs, geysers, and solfataras) provide a wide range of habitats well suited to thermophiles. The sites within the park are also relatively easy to access, especially compared to some other high temperature habitats, such as deep-sea hydrothermal vents. The application of molecular phylogenetic approaches to the study of Yellowstone's thermophilic microbial communities has revealed levels of diversity much higher than previously suspected. Octopus Spring's cyanobacterial mats had been extensively studied by traditional culture techniques, yet examination of 16S rRNA sequences yielded multiple sequences ("phylotypes") that did not match those of any organism that had been previously cultured from the spring (Ward et al., 1990; Weller et al., 1991). Additionally, the Octopus Spring pink filamentous bacterial community was characterized and it was shown that the dominant organism in that community is a member of the hydrogen-oxidizing *Aquificales* (Reysenbach et al., 1994). Phylogenetic analysis of the bacterial community of Obsidian Pool has also revealed a number of novel phylotypes and lineages (Hugenholtz et al., 1998). An unprecedented level of archaeal diversity was detected in Obsidian Pool (Barns et al., 1994). Based on the detection of phylotypes that were representative of an extremely deeply diverging archaeal lineage, a third kingdom in the *Archaea* was proposed, the "*Korarchaeota*" (Barns et al., 1996). These results point to a level of complexity in thermophilic microbial communities that is far greater than previously anticipated.

This study employs a molecular phylogenetic approach to examine the microbial diversity of a thermal stream flowing into Obsidian Pool in Yellowstone National Park. Four microbial community samples were taken along a temperature gradient in the stream and were analyzed by comparing isolated 16S rRNA sequences. We report here on the prevalence of the *Aquificales* in this community and on shared bacterial community members whose communities are associated with Obsidian Pool.

2. MATERIALS AND METHODS

2.1. Sample Collection

Samples were collected from a high temperature stream (pH 6.3) that flows into the Obsidian Pool thermal spring in Yellowstone National Park. The stream was colonized by visible black filamentous communities attached to the stream bed. Four samples of filaments were taken along a temperature gradient and fixed in 70% (w/w) ethanol. The temperatures of the samples were recorded as 66, 68, 70, and 76°C.

2.2. DNA Extraction

Community DNA was extracted by using a modification of the procedure described by Ausubel et al. (1988). The sample (1.5 mL) was thawed, spun in a microcentrifuge for 30 s, resuspended in 567 μL of Tris/EDTA TE, 38 μL of 10% SDS, 153 μL of 0.5 M EDTA, and 10 μL of 20 mg/ml Proteinase K, and was incubated for 1 h at 37°C. After 128 μL of 5 M NaCl and 102 μL of a 10% hexadecyltrimethyl ammonium bromide (CTAB) CTAB–0.7% NaCl solution were added, the sample was mixed thoroughly and incubated for 30 min at 65°C. The DNA was extracted once with an equal volume of phenol:chlorform:isoamyl alcohol (25:24:1), extracted with an equal volume of chloroform, and precipitated in 0.6 volume isopropanol. The DNA was washed with 70% ethanol, dried, and resuspended in 20 μL of TE buffer.

2.3. DNA Amplification

Bacterial 16S rRNA genes were amplified by PCR (Saiki et al., 1988) using a bacterial-specific forward primer corresponding to nucleotide positions 8 to 27 in *Escherichia coli* 16S rRNA gene (forward primer 8FPL: 5′-GCGGAATCCGCGGCCGCTGCAGAGTT TGATCCTGGCTCAG-3′) and a universal reverse primer corresponding to the complement of *E. coli* positions 1510 to 1492 (reverse primer 1492RPL: 5′-GGCTCGAGCGGC-CGCCCGGGTTACCTTGTTACGACTT-3′). An archaeal-specific forward primer (4F: 5′-TCCGGTTGATCCTGCCRG-3′) was also used in conjunction with the 1492RPL primer to amplify any archaeal 16S rRNA gene sequences in the sample. The reaction mixture consisted of 10 mM Tris (pH 8.3), 50 mM KCl, 1.5 mM $MgCl_2$, 200 μM dNTP's, 0.2% Igepal, 1 ng of each primer, 1 U of *Taq* polymerase, and approximately 50 ng of sample DNA. The reaction mixtures were incubated in a thermal cycler (Perkin-Elmer) for 5 min at 94°C and for 30 amplification cycles of 30 s at 94°C, 30 s at 55°C, and 1 min 30 s at 72°C, followed by 10 min at 72°C. Amplified DNA was gel purified using a Gene Clean DNA purification kit (Bio 101, Inc.).

2.4. Cloning and Sequencing

The amplified 16S rDNA fragments were cloned into pBluescript and transformed into competent cells of *E. coli* JM 109 (Promega). Clones that contained insert DNA of approximately 1500 base pairs were selected using the agarose gel screening procedure outlined by Sekar (1987). Forty insert-containing clones were randomly selected from each of the four samples, for a total of 160 clones screened. The clones were reamplified by using the same PCR conditions outlined before, substituting the primers with M13 lac Z forward and reverse primers (Perkin-Elmer), and raising the annealing temperature to 65°C. The reaction mixtures were incubated for 5 min at 94°C and subjected to 30 amplification cycles of 30 s at 94°C, 30 s at 65°C, and 1 min and 30 s at 72°C, followed by 10 min at 72°C.

Different clone types (phylotypes) were identified using a double digest with *Hin* P1 and *Msp* 1 and restriction fragment length polymorphism (RFLP) analysis. The PCR product from each clone (14.5 μL) was mixed with 2 μL NEB 2 buffer, 2 μL 0.1% Triton X-100, 1 μL *Hin* P1 endonuclease (10U/μL), and 0.5 μL *Msp* 1 endonuclease (20U/μL). The digests were electrophoresed in a 3.5% Nusieve gel in TBE buffer at 80 V. Clones that had different banding patterns, indicative of different phylotypes, were selected for sequencing. Selected clones were cycle sequenced using a reaction mixture of 4 μL FS sequenase (Perkin-Elmer), 2 μL purified PCR product, 3 μL sterile H_2O, and 1.6 pmol primer. Both strands of each clone were fully sequenced using each of the following primers: 8F, 515F, 906F, 519R, 907R, and 1492R. Sequences were run on an Applied Biosystems 377 sequencing gel. Compiled sequences were all approximately 1450 nucleotides long, representing nearly complete 16S rRNA gene sequences.

2.5. Phylogenetic Analysis

16S rRNA sequences were aligned with a subset of bacterial and archaeal 16S rRNA sequences provided by the Ribosomal Database Project (Larsen et al., 1993) and Genbank. Sequence secondary structure (Gutell, 1993) and conserved sequence regions (Ward et al., 1987) were used as guides for the alignment. Approximately 1350 nucleotide positions were included in the analysis. Distance trees were generated using the least-squares algorithm (DeSoete, 1983) with the Jukes and Cantor correction (1969). Maximum likelihood trees were constructed (Felsenstein, 1981) using the fast DNAml program (Larsen et al., 1993). The sequences were submitted to Genbank as AF018186–AF018200.

3. RESULTS

3.1. Cloning and Sequence Analysis

DNA isolated from the Obsidian Pool stream samples was used to amplify the microbial community 16S rDNA using bacterial-specific primers. Attempts to amplify archaeal 16S rRNA sequences using Archaea-specific primers were unsuccessful. A forty-clone library from each sample was screened by RFLP analysis to identify unique insert-containing clones. A total of 18 different RFLP banding patterns was observed in the 160 clones screened (see Fig. 1). Table 1 is a summary of the phylotypes observed in each sample. Clone OPS1 was the most frequently observed pattern, comprising 55.6% of all sampled clones. Clone OPS3 was the second most common (23.1% of all clones analyzed). OPS12 represented 4.38%, and OPS6 represented 3.75% of the clones. All other clone types represented less than 2% of the total sample. Additionally, different percentages of the clones were observed in the different temperature samples (Table 1). These percentages reflect the relative abundance of these phylotypes in the clone library but do not necessarily reflect *in situ* distributions of the phylotypes. One chimeric type of clone (between OPS1

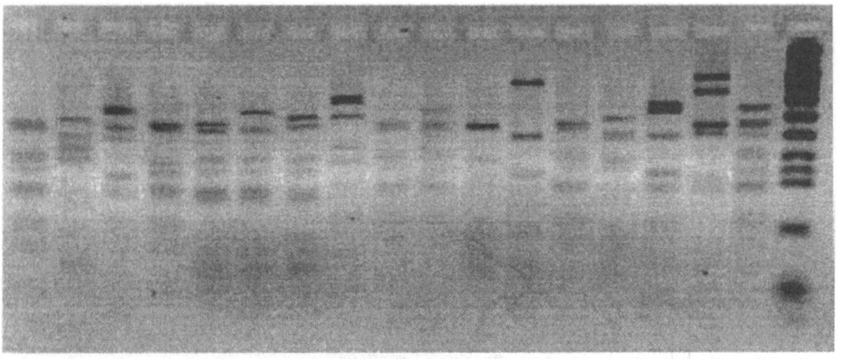

Figure 1. Composite image of restriction fragment length polymorphism (RFLP) patterns for Obsidian Stream phylotypes.

and OPS5) was identified (OPS6) at all temperatures (Hugenholtz, pers. comm.). Because the identical clone was obtained from four different samples, it is perhaps questionable whether this is indeed a chimera.

3.2. Phylogenetic Analysis

Analysis using the DeSoete algorithm for distance-based phylogenetic trees and fast DNAml to generate phylogenetic trees resulted in the same branching patterns, lending

Table 1
Percentage Abundance of Clone Types ("Phylotypes") from RFLP Pattern (See Methods)

Phylotype	Group of closest affiliation	66°C (%)	68°C (%)	70°C (%)	76°C (%)
OPS1	Black filaments	90	30	27.5	75
OPS2	C/F/B/ group	2.5	—	—	2.5
OPS3	*Planctomycetales*	5	42.5	45	—
OPS4	Black filaments	2.5	—	—	—
OPS5	*Aquificales*	—	—	—	5
OPS7	Black filaments	—	—	2.5	2.5
OPS8	*Thermotogales*	—	—	2.5	2.5
OPS9	Black filaments	—	—	—	2.5
OPS10	Black filaments	—	—	2.5	2.5
OPS11	Black filaments	—	—	—	2.5
OPS12	Green nonsulfur bacteria	—	12.5	5	—
OPS13	Black filaments	—	7.5	—	—
OPS14	Black filaments	—	2.5	—	—
OPS15	*Thermus*	—	—	2.5	—
OPS16	*Planctomycetales*	—	—	2.5	—
OPS17	C/F/B group	—	—	2.5	—
OPS18	Black filaments	—	2.5	—	—

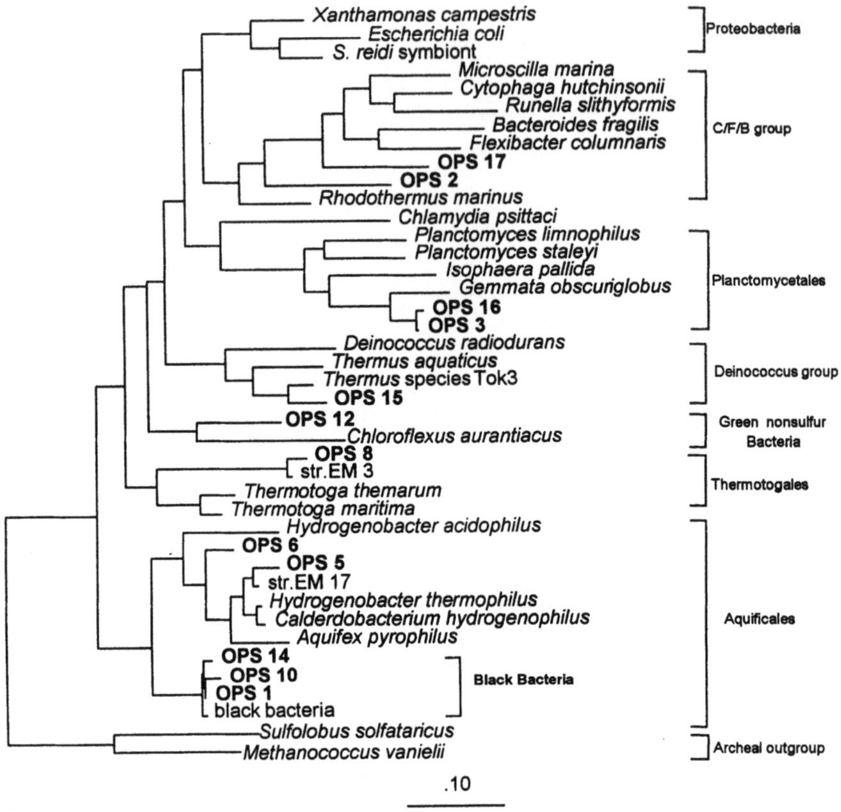

Figure 2. Maximum likelihood tree of major phylotypes (OPS) found along a temperature gradient in the Obsidian Pool stream. The scale bar represents the expected number of changes per sequence position.

confidence to the proposed evolutionary relationships (see Fig. 2). The analyzed sequences fall into several distinct groups within the bacterial domain.

Maximum likelihood analysis placed a number of similar sequences in a group with an as yet uncultivated organism referred to as "black filaments," which were first detected in Yellowstone's Calcite Springs (Reysenbach et al., 1995) and subsequently in the Obsidian Pool (OPB13; Hugenholtz et al., 1998). OPS 1, 4, 7, 9, 10, 11, 13, 14, and 18 all grouped with this organism and differed in sequence by less than 1%. The sequences within this clade are approximately 80% similar to the sequences of the *Aquificales*, suggesting that they represent, at the very least, a new genus within the *Aquificales*, and perhaps a new order within the domain Bacteria. OPS5 is closely related to the *Aquificales* genera *Hydrogenobacter*, *Calderobacterium*, and *Aquifex* (Pitulle et al., 1994) and the phylotype of the pink filaments from Octopus Spring (EM 17; Reysenbach et al., 1994). OPS8 forms a clade with EM 3, an as yet uncultivated organism that falls within the order *Thermotogales* (Reysenbach et al., 1994). OPS15 is most closely related to the genus *Thermus*. OPS12 groups closest to the family Chloroflexaceae, a group of anaerobic phototrophs, and is identical to the phylotype OPB65 (Hugenholtz et al., 1998). OPS3 and OPS16 are members of the order

Planctomycetales (Ward et al., 1995) and are most closely related to the organisms *Gemmata obscuriglobus* and *Isophaera pallida*. OPS2 and OPS17 fall closest to *Rhodothermus marinus*, a thermophilic organism near the root of the *Cytophaga-Flexibacter-Bacteroides* (C-F-B) group (Andresson et al., 1994). OPS2 is very similar in sequence to OPB88 which was obtained from the Obsidian Pool (Hugenholtz et al., 1998). OPS6 is a chimeric molecule of OPS1 and OPS5.

4. DISCUSSION

Phylogenetic analysis of the diversity of the communities along the temperature gradient of a thermal stream that flows into the Obsidian Pool revealed relatively low diversity (eighteen distinct phylotypes of 160 clones screened) compared to community diversity observed in the pool itself (Hugenholtz et al., 1998). Half of those clones were members of the "black filamentous" clade, first reported from Calcite Springs (Reysenbach et al., 1995) and most closely related to the Aquificales. This group is geographically widespread, and members have been sequenced from geothermal springs in Japan (GAN and NAK9; Yamamoto et al., 1998), the Azores (AZ4), deep-sea hydrothermal vents (VC-39), and Mammoth Hot Springs, Yellowstone National Park (TTG5; Reysenbach, manuscript in prep). The communities may appear black, gray, gray-green, yellow or white, most likely depending on the minerals precipitated on their cell walls. They all thrive in near neutral environments (pH 6.0–8.3) at temperatures from about 60°C to about 86°C. Based on the clone library percentages and preliminary *in situ* hybridization studies using fluorescent oligonucleotide probes specific for members of this filamentous clade (data not shown), these organisms are the dominant community members. Although the communities are filamentous, the individual cells are rods of approximately 5–10 μm long. In the Obsidian Pool, these organisms appear black and can be seen attached to debris such as grass on the edges of the pool. In the stream flowing into the Obsidian Pool (this study), the organisms form very defined, thick (several millimeters) mats. The relatively low diversity of these communities makes them ideal models for studying the physiological ecology of thermophiles.

The phylotypes OPS1, OPS4, OPS7, OPS9, OPS10, OPS11, OPS13, OPS14, and OPS18 are all members of the "black filament" clade. The 16S rRNA sequences differ by less than 1%. This molecular microdiversity may reflect a preference for a specific ecotype, perhaps determined by temperature, that enables the entire population of black filaments to thrive over a broader temperature regime. Similar microheterogeneity has been seen in many other studies of environmental microbial diversity (for example, Angert et al., 1993; DeLong et al., 1998; Hugenholtz et al., 1998, Ward et al., this volume). It has been specifically shown that the molecular microdiversity seen within populations of *Prochlorococcus* reflects multiple ecotypes (Moore et al., 1998). Overall, OPS1 was the most frequently observed phylotype and represented the dominant phylotype in the 66°C and 76°C samples. The other phylotypes associated with the black filamentous group were relatively minor community constituents that occurred in all samples (see Table 1).

The sequence of OPS5 grouped within the order *Aquificales*, the extremely thermophilic bacterial lineage (Burggraf et al., 1992), related to the black filamentous phylotypes. OPS5 was most closely affiliated with EM 17, a phylotype detected in Yellowstone's

Octopus Spring outflow and identified conclusively by *in situ* hybridization as the pink filamentous morphotype (Reysenbach et al., 1994). This phylogenetic data contributed to the successful cultivation of this organism, now named *Thermocrinis ruber* (Huber et al., 1998). The *Aquificales* are typically chemolithotrophic hydrogen oxidizers and are extreme thermophiles whose temperature optima are above 70°C (Huber et al., 1992; Igarashi et al., 1990). OPS5 appeared only in the 76°C community.

OPS8 formed a clade with the uncultivated phylotype EM 3, another organism detected in the Octopus Spring outflow (Reysenbach et al., 1994). EM 3 is a deeply branching member of the order *Thermotogales*, a group of thermophilic organoheterotrophs. The OPS8 phylotype was detected at both 70 and 76°C.

The phylotype OPS3 fell within the order *Planctomycetales* and is most closely affiliated with *Gemmata obscuriglobus* (Franzmann et al., 1984). The Planctomycetes are a widely distributed group of chemoheterotrophic budding bacteria that lack the peptidoglycan cell wall otherwise common to all bacteria (Fuerst, 1995). *Isophaera pallida*, another related planctomycete, is a frequent member of North American hot spring cyanobacterial mat communities (Giovannoni et al., 1987) and has been detected in other Yellowstone thermal features (Ward et al., 1987). OPS3 was observed at 66, 68, and 70°C, and was the dominant phylotype at the two latter temperatures (see Table 1). OPS16, represented by a single observed clone in the 70°C sample, also fell into the Planctomycetes group and formed a clade with OPS3. Although this phylotype is abundant in this study, it does not necessarily reflect the *in situ* abundance of this organism. Futher *in situ* hybridization studies would be necessary to evaluate the actual abundance of *Planctomycetales* in the thermal stream.

A number of the other phylotypes within the stream were also observed in the much more diverse Obsidian Pool community (Hugenholtz et al., 1998). For example, OPS12 is identical in sequence to OPB65, and the OPS2 phylotype was shared with OPB88 from the Obsidian Pool (Hugenholtz et al., 1998). OPS12 was observed in the 68°C and 70°C communities and grouped with *Chloroflexus aurantiacus*, an anaerobic phototrophic bacterium. The family *Chloroflexaceae* is characterized by a filamentous morphology, gliding motility, and the use of hydrogen sulfide as an electron donor. *C. aurantiacus* is widely distributed in hot springs throughout the world and is a frequent constituent of high temperature microbial communities in Yellowstone (Brock, 1978). The OPS2 and OPB88 sequences fall into the C-F-B group and are closely associated with *Rhodothermus marinus*, a heterotrophic thermophile previously isolated from Icelandic marine hot springs (Alfredsson et al., 1988). *Rhodothermus marinus* grows optimally in a temperature range of 65–75°C, which correlates with the occurrence of OPS2 in both the 66°C and 76°C communities. OPS17 also grouped with the C-F-B group, and was represented by only one clone in the 70°C community (Table 1).

The similarities between the bacterial diversity within the thermal stream at the Obsidian Pool and within the pool itself is not surprising, but rather encouraging. The thermal spring offers a good model for studying the interaction of these communities in a system of relatively low diversity. The Obsidian Pool environment, however, is less tangible because it is a highly turbulent, mixed environment, and temperatures vary considerably within very short distances. Furthermore, the prevalence of the 'black filamentous' clade in nearly neutral thermophilic communities requires further study, and this thermal stream offers a cell-characterized environment to attempt such a study. Understanding the contribution of this group to the productivity of high temperature ecosystems is currently under investigation in our laboratory. Preliminary geochemical comparisons between Calcite Springs and Obsidian Pool suggests that there are significant differences in the geochemical milieu in which these organisms can thrive, indicative perhaps of a physiologically diverse organism. That this lineage and the Aquificales represent one of the most deeply rooted branches in the bacterial 16S rRNA phylogenetic tree may suggest that they are good analogs for studying the early evolution of metabolic cycles. The recently available genome of *Aquifex aeolicus* will facilitate this type of investigation (Deckert et al., 1998).

The recent studies of the microbial diversity and ecology of thermal springs in Yellowstone National Park were laid on the foundations of research established by Brock and colleagues. However, the advances in molecular approaches to microbial ecology have enabled researchers to revisit and expand these foundations. Consequently, Yellowstone's thermal springs will remain excellent natural laboratories for understanding life at high temperatures.

REFERENCES

Adams, M. W. W., Perler, F. B., and Kelly, R. M. 1995. Extremozymes—expanding the limits of biocatalysis. *Biotechnology* **13**:662–668.

Alfredsson, G., Kristjansson, J., Hjrleifsdottir, S., and Stetter, K. 1988. *Rhodothermus marinus*, gen. nov., sp. nov., a thermophilic, halophilic bacterium from submarine hot springs in Iceland. *J. Gen. Microbiol.* **134**:206–299.

Amann, R., Ludwig, W., and Schleifer, K. 1995. Phylogenetic identification and in situ detection of individual microbial cells without cultivation. *Microbiol. Rev.* **59**:143–169.

Andresson, O., and Fridjonsson, O. 1994. The sequence of the single 16S rRNA of the thermophilic Eubacerium *Rhodothermus marinus* reveals a distant relationship to the group containing *Flexibacter*, *Bacteroides*, and *Cytophaga* species. *J. Bacteriol.* **176**:6165–6169.

Angert, E. R., Clements, K. D., and Pace, N. R. 1993. The largest bacterium. *Nature* **362**:239–241.

Ausubel, F., Brent, R., Kingston, R., Moore, D., Seidman, J. G., Smith, J., and Struhl, K. (eds). 1988. *Current protocols in molecular biology*. New York: Greene Publishing and Wiley Interscience.

Barns, S., Delwiche, C., Palmer, J., and Pace, N. 1996. Perspectives on archaeal diversity, thermophily and monopyly from environmental rRNA sequences. *Proc. Natl. Acad. Sci. USA* **93**:9188–9193.

Barns, S., Fundyga, R., Jeffries, M., and Pace, N. 1994. Remarkable archaeal diversity detected in a Yellowstone National Park hot spring environment. *Proc. Natl. Acad. Sci. USA* **91**:1609–1613.

Blochl, E., Rachel, R., Burggraf, S., Hafenbradl, D., Jannasch, H. W., and Stetter, K. O. 1997. *Pyrolobus fumarii*, gen. and sp. nov., represents a novel group of archaea, extending the upper temperature limit for life to 113°C. *Extremophiles* **1**:14–21.

Brock, T. D. 1985. Life at high temperatures. *Science* **230**:132–138.

Brock, T. D. 1987. The study of microorganisms in situ: Progress and problems. *Symp. Soc. Gen. Microbiol.* **41**: 1–17.

Brock, T. D. 1978. *Thermophilic microorganisms and life at high temperatures*. New York: Springer-Verlag.

Burgraff, S., Olsen, G., Stetter, K., and Woese, C. 1992. A phylogenetic analysis of *Aquifex pyrophilus*. *Syst. Appl. Microbiol.* **15**:352–356.

Corliss, J. 1990. Hot springs and the origin of life. *Nature* **347**:624.

Deckert, G., Warren, P. V., Gasterland, T., Young, W. G., Lennox, A. L., Graham, D. E., Overbeek, R., Snead, M. A., Keller, M., Aujay, M., Huber, R., Feldman, R. A., Short, J. M., Olsen, G. J., and Swanson, R. V. 1998. The complete genome sequence of the hyperthermophilic bacterium *Aquifex aeolicus*. *Nature* **392**:353–358.

DeLong, E., Wickham, G., and Pace, N. 1989. Phylogenetic stains: Ribosomal RNS-based probes for the identification of single cells. *Science* **243**:1330–1363.

DeSoete, G. 1983. A least squares algorithm for fitting additive trees to proximity data. *Psycometrika* **48**:621–626.

Felsenstein, J. 1981. Evolutionary trees from DNA sequences: A maximum likelihood approach. *J. Mol. Evol.* **17**:368–376.

Fox, G., Stackebradt, E., Hespell, R., Gibson, J., Manlihoff, J., Dyer, T., Wolfe, R., Balch, W., Tanner, R., Margum, L., Zablen, L., Blakemore, R., Gupta, R., Bonen, L., Lewis, B., Stahl, D., Lueshresen, K., Chen, K., and Woese, C. 1980. The phylogeny of prokaryotes. *Science* **209**:457–463.

Franzmann, P. D., and Skerman, V. B. 1984. *Gemmata obscuriglobus*, a new genus and species of budding bacteria. *Antonie Leeuwenhoek*, **50**:261–268.

Fuerst, J. 1995. The planctomycetes: Emerging models for microbial ecology, evolution, and cell biology, *Microbiol.* **141**:1493–1506.

Giovannoni, S. J., Schabtach, E., and Castenholz, R. W. 1987. *Isophaera pallida*, gen, and comb, nov., a gliding, budding eubacterium from hot springs. *Arch. Microbiol.* **147**:276–284.

Gutell, R. 1993. The collection of small subunit (16S and 16S like) ribosomal RNA structures. *Nucleic Acid Res.* **21**:3051–3054.

Huber, R., Wilharm, T., Huber, D., Trincone, A., Burggraf, S., Konig, H., Rachel, R., Rockinger, I., Frike, H., and Stetter, K. 1992. *Aquifex pyrophilus* gen. sp. nov., represents a novel group of marine hyperthermophilic hydrogen-oxidizing bacteria. *Syst. Appl. Microbiol.* **15**:340–351.

Huber, R., Eder, W., Heldwein, S., Wanner, G., Huber, H., Rachel, R., and Stetter, K. O. 1998. *Thermocrinus ruber* gen. nov., sp. nov., a pink filament-forming hyperthermophilic bacterium isolated from Yellowstone National Park. *Appl. Environ. Microbiol.* **64**:3576–3583.

Hugenholtz, P., Pitulle, C., Hershberger, K. L., and Pace, N. R. 1998. Novel division level bacterial diversity in a Yellowstone hot spring. *J. Bacteriol.* **180**:366–376.

Igarashi, Y., and Kodama, T. 1990. *Hydrogenobacter thermophilus*: Its unusual physiological properties and phylogenetic position in the microbial world. *FEMS Microbiol. Rev.* **87**:403–406.

Jukes, T., and Cantor, C. 1969. Evolution of protein molecules, In Munro, H. (ed.), *Mammalian protein metabolism* (pp. 21–132). New York: Academic Press.

Larsen, N., Olsen, G., Maidak, B., McCaughey, N., Overbeek, R., Macke, T., Marsh, T., and Woese, C. 1993. The Ribosomal Database Project. *Nucleic Acids Res.* **21**:3021–3023.

Moore, L. R., Rocap, G., and Chisholm, S. W. 1998. Physiology and molecular phylogeny of coexisting *Procholococcus* ecotypes. *Nature* **393**:464–467.

Olsen, G., Woese, C., and Overbeek, R. 1994. The wind of (evolutionary) change: Breathing new life into microbiology. *J. Bacteriol.* **176**:1–6.

Pitulle, C., Yang, Y., Marchani, M., Moore, E., Seifert, J., Aragno, M., Jurtshuk, P., Jr., and Fox, G. 1994. Phylogenetic position of the genus *Hydrogenobacter*. *Int. J. Syst. Bacteriol.* **44**:620–626.

Reysenbach, A.-L., Ehringer, M. A., Hershberger, K., Favre, R., Cermola, M., Palsdottir, A., and Eggertsson, G. 1995. Analysis of microbial hot spring communities from Yellowstone National Park and Iceland by 16S rRNA sequence analysis. *Environment* **39**:5–6.

Reysenbach, A.-L., Wickham, G., and Pace, N. 1994. Phylogenetic analysis of the hyperthermophilic pink filament community in Octopus Spring, Yellowstone Natl. Park. *Appl. Environ. Microbiol.* **60**:2114–2119.

Reysenbach, A.-L., Wickham, G., and Pace, N. 1992. Differential amplification of rRNA genes by polymerase chain reaction. *Appl. Env. Microbiol.* **58**:3417–3418.

Saiki, R., Gelfand, D., Stoffel, S., Scharf, S., Higuchi, R., Horn, G., Mullis, K., and Erlich, H. 1988. Primer-directed enzymatic amplification of DNA with a thermostable DNA polymerase. *Science* **239**:487–491.

Sekar, V. 1987. A rapid screening procedure for the identification of recombinant bacterial clones. *BioTechniques* **5**:11–13.

Schonheit, P., and Schafer, T. 1995. Metabolism of hyperthermophiles. *World J. Microbiol. Biotech.* **11**:26–57.

Stetter, K., Fiala, G., Huber, G., and Segerer, A. 1990. Hyperthermophilic microorganisms. *FEMS Microbiol.* **75**:117–124.

Tilman, D., and Downing, J. A. 1994. Biodiversity and stability in grasslands. *Nature* **367**:363–365.

Tilman, D., Knops, J., Wedin, D., Reich, P., Ritchie, M., and Seimann, E. 1997. The influence of functional diversity and composition on ecosystem processes. *Science* **277**:1300–1302.

Ward, D. M., Bateson, M., Weller, R., and Ruff-Roberts, A. 1992. Ribosomal RNA analysis of microorganisms as they occur in nature. *Adv. Microbiol. Ecol.* **12**:219–286.

Ward, D. M., Weller, R., and Bateson, M. 1990. 16S rRNA sequences reveal numerous uncultured microorganisms in a natural community. *Nature* **345**:63–65.

Ward, D. M., Tayne, T., Anderson, K., and Bateson, M. 1987. Community structure and interactions among community members in hot springs cyanobacterial mats. *Symp. Gen. Soc. Microbiol.* **41**:179–210.

Ward, N., Rainey, F., Stackbrandt, E., and Schesner, H. 1995. Unraveling the extent of diversity within the order *Planctomycetales*. *Appl. Env. Microbiol.* **61:**2270–2275.

Weller, R., Walsh, J., and Ward, D. 1991. 16S rRNA sequences of uncultivated hot spring cyanobacterial mat inhabitants retrieved as randomly primed cDNA. *Appl. Environ. Microbiol.* **57:**1145–1151.

Woese, C., Kandler, O., and Wheelis, M. 1990. Towards a natural system of organisms: Proposal for the domains Archaea, Bacteria, and Eucarya. *Proc. Natl. Acad. Sci. USA* **87:**4576–4579.

Woese, C. 1987. Bacterial evolution. *Microbiol. Rev.* **51:**221–227.

Yamamoto, H., Hirashi, A., Kato, K., Chiura, H. X., Maki, Y., and Shimizu, A. 1998. Phylogenetic evidence for the existence of novel thermophilic bacteria in hot spring sulfur-turf microbial mats in Japan. *Appl. Environ. Microbiol.* **64:**1680–1687.

8

Isolation of Hyperthermophilic *Archaea* Previously Detected by Sequencing rDNA Directly from the Environment

Siegfried Burggraf, Robert Huber, Thomas Mayer, Petra Rossnagel, and Reinhard Rachel

1. INTRODUCTION

Hydrothermal systems are an exciting source of organisms (Blöchl et al., 1995; Stetter, 1992). To date, an unexpectedly large number of unique phenotypes, almost exclusively *Archaea*, have been isolated from biotypes at a temperature above 85°C, and it seems very likely that the limitations of the conventional enrichment culturing methods have prevented us from assessing the true extent of this number. The first indication of the real variety of *Archaea* living in hydrothermal ecosystems was given by analyses of 16S rDNA sequences obtained directly without previous cultivation from samples from the Obsidian Pool, a hot spring at Yellowstone National Park. A large number of archaeal sequences could be detected, all of which differed from those of known species (Barns et al., 1994a,b). Knowing the 16S rRNA sequences permits designing specific oligonucleotide probes that can be used in whole cell hybridization experiments (Amann et al., 1990b; Burggraf et al., 1994; DeLong et al., 1989; Stahl and Amann, 1991). This method allows determining the morphotype of an organism that corresponds to a new sequence, but the examination of its physiological and biochemical properties still requires a growing culture. For more than a century, Koch's plating technique was the fundamental means to obtain pure cultures of

Siegfried Burggraf, Robert Huber, Thomas Mayer, Petra Rossnagel, and Reinhard Rachel • Lehrstuhl für Mikrobiologie and Archaeenzentrum, Universität Regensburg, 93053 Regensburg, Germany.

Thermophiles: Biodiversity, Ecology, and Evolution, edited by Reysenbach *et al.* Kluwer Academic / Plenum Publishers, New York, 2001.

microorganisms (Koch, 1881). However, by this technique, only a small percentage of organisms that form colonies can be obtained in pure culture. Serial dilutions, another isolation technique, works only for organisms that are predominant within environmental samples. A revolutionary new approach is the use of a strongly focused infrared laser beam ("optical tweezers") to separate single microorganisms (Ashkin and Dziedzic, 1987; Ashkin et al., 1987; Huber et al., 1995). This chapter describes the isolation of three organisms previously detected only by their 16S rDNA sequences by a procedure that combines visual recognition of single cells in enrichment cultures by phylogenetic staining (DeLong et al., 1989; Kane et al., 1993) and cloning by "optical tweezers" (Huber et al., 1995).

2. MATERIALS AND METHODS

2.1. Sampling and Enrichment of Themophiles

Aerobic and anaerobic water and sediment samples were taken from the Obsidian Pool, a hot spring located in the Mud Volcano area of Yellowstone National Park, Wyoming (Barns et al., 1994a). The samples were collected at temperatures between 73°C and 93°C and a pH of 6.7 and consisted of black sandy material, obsidian and possibly iron sulfide. Samples were collected in 100-mL storage bottles and tightly sealed by rubber stoppers. Immediately after collection, solutions of sodium sulfide and sodium dithionite were injected into about half of the samples to reduce oxygen (Stetter 1982). The other half of the samples remained aerobic. The samples were transported to the laboratory at room temperature. For enrichment culturing, the anaerobic technique of Balch and Wolfe was followed (Balch and Wolfe 1976). Cultivation media at a pH of 7 and an ionic strength adjusted to that of the original samples were used. The media contained different possible energy sources (Balch et al., 1979; Huber et al., 1989; Huber and Stetter, 1992; Stetter, 1982; Stetter et al., 1981, 1987). The different cultivation media were inoculated with the aerobic or anaerobic sample material (10% inoculum) and incubated at three different temperatures (73, 83 and 93°C). Growth was observed by microscopic inspection.

2.2. Cell Fixation

Aliquots of the cultures (3 mL) were fixed by adding three volumes of a paraformaldehyde solution [4% w/v in phosphate-buffered saline (PBS)] directly to the culture. After the fixation (3h at room temperature) cells were washed with PBS and stored it a 1:1 mixture of PBS and 98% ethanol at –20°C (Amann et al., 1990a). Fixed cells were spotted on precleaned, gelatin-coated [0.1% gelatin, 0.01% $KCr(SO_4)_2$] microscope slides (Paul Marienfeld KG, Bad Mergentheim, Germany), dried at 46°C for 30min, and dehydrated in 50, 80, and 98% (v/v) ethanol (3 min each).

2.3. Whole Cell Hybridization

Whole cell hybridization with fluorescently labeled oligonucleotide probes were performed as described (Amann et al., 1990b; Burggraf et al., 1994). Probes used to distinguish between *Bacteria* and *Archaea* were EUB338 (Amann et al., 1990a) and ARCH915 (Stahl and Amann, 1991), respectively. A probe specific for *Desulforococcus mobilis* and sequences obtained by sequencing of ribosomal DNA PCR-amplified directly from the Obsidian Pool samples [sequence pSL91 (Barns et al., 1994b) and sequence pJP74 (Barns et al., 1994a)] had the sequence 5′ TAT GGG GGA TTA GCA CCA 3′ and corresponded to positions 164 to 181 in the 16S rRNA [*E. coli* numbering (Brosius et al., 1981)]. The oligonucleotides were synthesized with a C6-TFA aminolink [6-(trifluoroacetyl-amino)-hexyl-(2-cyanoethyl)-(*N,N*-diisopropyl)-phosphoramidite] at the 5′-end by MWG Biotech (Ebersberg, Germany). Labeling of the oligonucleotides with tetramethylrhodamine-5-isothiocyanate (TRITC, Research Organics, Cleveland, USA) or 5(6)-carboxyfluorescein-*N*-hydroxysuccinimide-ester (FLUOS, Boehringer Mannheim, Germany) and purification was performed as previously described (Amann et al., 1990b).

2.4. Fluorescence Microscopy

A Nikon Microphot EPI-FL microscope equipped with a UV lamp and the filter sets G-1B and B-2A (Nikon) was used to visualize the hybridized cells.

2.5. Cell Separation

Cells were isolated by optical trapping in a laser beam using a modified computer-controlled inverse microscope (Axiovert IM35, Zeiss, Oberkochen, Germany) equipped with an oil immersion objective (100/1.3) and a Nd:YAG laser (wavelength: 1064 nm; maximum power: 1 watt; ADLAS, Lübeck, Germany) (Huber et al., 1995). After separation, single cells were flushed into a fresh anaerobic medium and were incubated at 83°C.

2.6. Sequencing of 16S rDNA

16S rDNA was amplified by PCR as described (Burggraf et al., 1997b). The PCR products were sequenced directly using an AmpliCycle Sequencing Kit (Perkin Elmer) and a set of standard archaeal "forward" and "reverse" primers (Burggraf et al., 1997b).

3. RESULTS

3.1. Whole Cell Hybridization

Within about 2 weeks of incubation at the different temperatures, a total of seventy-five enrichment cultures were obtained. Microscopic inspection showed morphologically heterogeneous cocci, different rods, and filaments. Hybridization with domain-specific, fluorescently labeled oligonucleotide probes (Amann et al., 1990a; Stahl and Amann 1991) revealed the distribution of *Archaea* and *Bacteria* in the different enrichment cultures (Fig. 1). Most of the cultures grown at 73°C contained only Bacteria. Only archaeal cells could be detected in the majority of the cultures at 83°C (Fig. 1). Our isolation attempts were focused on cultures that contained mainly *Archaea* to cultivate some of the new *Archaea* corresponding to the previously detected rDNA sequences that Barns et al. obtained directly from the environment (Barns et al., 1994a,b). In an 83°C morphologically diverse

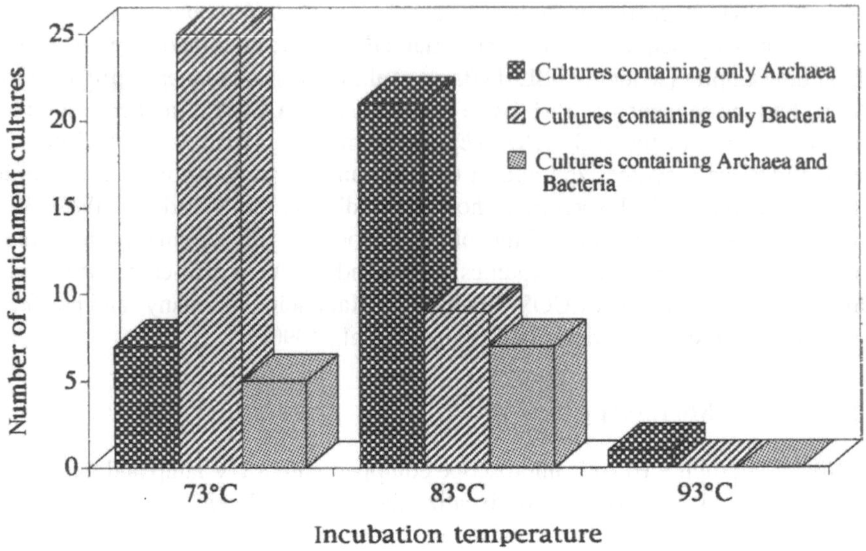

Figure 1. Distribution of *Archaea* and Bacteria in the different enrichment cultures.

enrichment culture, we focused on grape-forming aggregates of cocci that exhibited positive hybridization with the probe specific for *Desulfurococcus mobilis* and the sequence type pSL91.

3.2. Isolation of Different Morphotypes

Purification of the grape-forming aggregates of cocci by conventional methods like plating or serial dilutions was impossible due to the low final cell titers (up to 1×10^7/mL) and the faster growth of *Thermoproteus*- and *Thermofilum*-like cells in the enrichments. Therefore we used "optical tweezers" (Ashkin and Dziedzic, 1987; Ashkin et al., 1987) to separate cells from the mixed population. By this micromanipulation technique, it was possible to separate a single cell from the grape-like aggregates directly under the microscope and to obtain a pure culture (designated M11TL) (Huber et al., 1995). Using "optical tweezers," two other distinct morphotypes were isolated and cultivated (cultures designated S10L and S10TFL) from mixed cultures that contained mainly *Archaea*, as determined by whole cell hybridization with the archaeal probe.

3.3. Phylogeny and Physiology

To determine the phylogenetic position of the new isolates, their 16S rDNAs were sequenced and aligned to a set of representative cultivated *Archaea* (Maidak et al., 1996) and to sequences directly from the environment (Barns et al., 1994a,b). As suggested by whole cell hybridization with the specific probe, the sequence of the novel isolate M11TL was identical to the sequence pSL91. The closest cultivated relative of M11TL was *Desulfurococcus mobilis* (Fig. 2). Isolate M11TL is a coccoid organism that forms grape-like

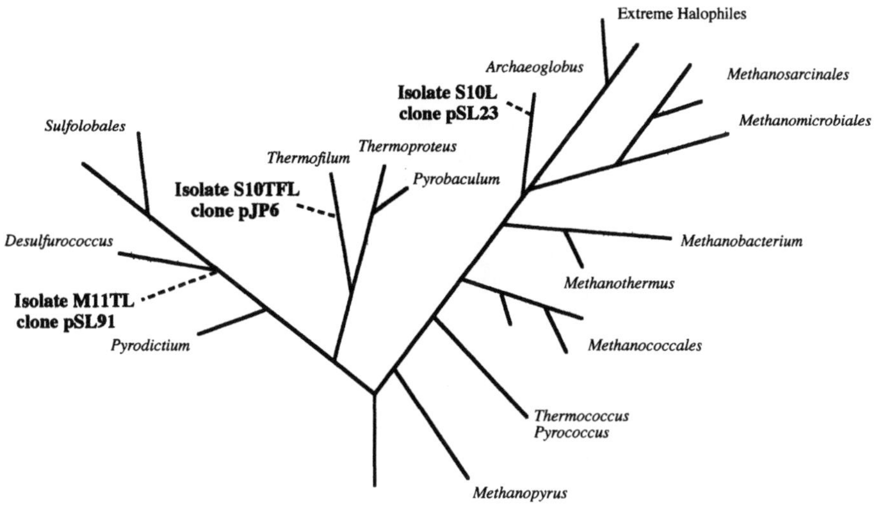

Figure 2. Phylogenetic tree of the domain *Archaea* (schematic drawing).

aggregates (Fig. 3-1). Based on physiological and phylogenetic analyses, it represents a new genus within the *Crenarchaeota* and is described as *Thermosphaera aggregans* (Huber et al., 1998). Isolates S10L and S10TFL were identical to sequence types pSL23 and pJP6, respectively (Fig. 2). These sequence types had also been detected earlier directly in Obsidian Pool samples by Barns et al. (Barns et al., 1994a,b). *Archaeoglobus fulgidus* and *Thermofilum pendens* were the closest cultivated relatives of S10L and S10TFL, respectively (Fig. 2). Cells of the isolate S10TFL are filamentous rods (Fig. 3-2). M11TL and S10TFL grow in low salinity by fermentation of complex organic material. Their upper temperature limit of growth is 90°C. Isolate S10L (Fig. 3-3) is a coccoid organism and probably represents a new species of the genus *Archaeoglobus*. It gains energy by reducing sulfate and in contrast to previously known members of the genus *Archaeoglobus*, S10L grows only at low salt concentrations up to 0.7% NaCl and therefore represents the first hyperthermophilic sulfate reducer from a nonmarine biotope.

4. SUMMARY

The isolation of three hyperthermophilic *Archaea* demonstrates the application of molecular techniques to enhance culturing strategies to examine the physiology and ecology of recently discovered novel organisms. The strategy is summarized in Fig. 4: (A) detection of novel 16S rDNA sequence types ("phylotypes") in environmental samples (Barns et al., 1994a,b), (B) screening enrichment cultures by whole cell hybridization, (C) isolation of identified single cells using "optical tweezers" and cultivation, and (D) confirming the phylogenetic affiliation by 16S rRNA sequence comparison. Our results validate the PCR data on the existence of a large archaeal community in a Yellowstone hot spring and also combine the study of pure cultures by molecular techniques for detection and quantification of microorganisms to provide basic information for a better understand-

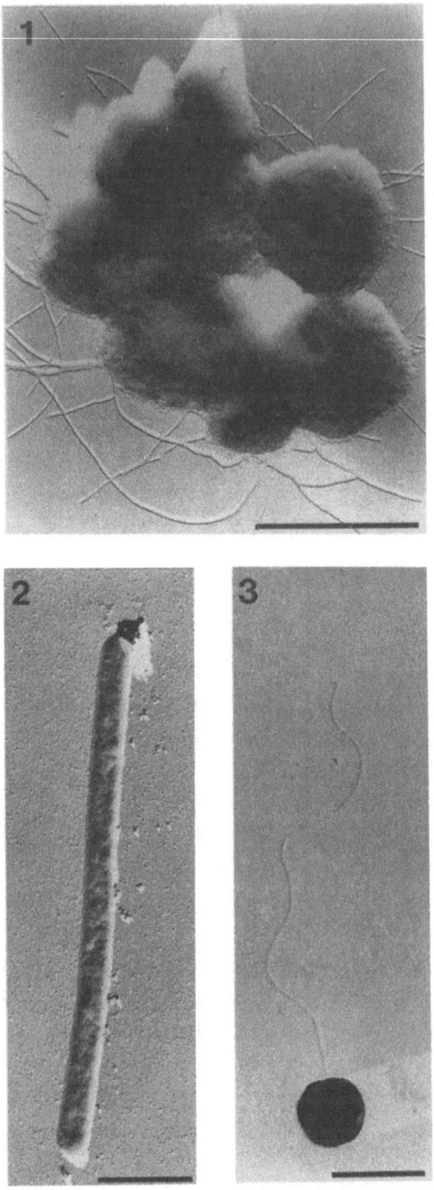

Figure 3. Electron micrographs of platinum-shadowed cells of isolate M11TL (1), isolate S10TFL (2), and isolate S10L (3). Scale bars: 1μm

ing of this microbial ecosystem. Our strategy should also be useful in isolating some of the many discovered, uncultivated organisms that occur in different biotopes (Amann et al., 1991; Barns et al., 1994a, b; DeLong, 1992; DeLong et al., 1994; Fuhrman et al., 1992; Giovannoni et al., 1990; Hugenholtz et al., 1988; Ward et al., 1990). Recently, a member of the *Korarchaeota*, a novel archaeal group (Barns et al., 1996) thus far only identified by its 16S rDNA sequence, was grown in mixed culture and its morphology was identified by *in*

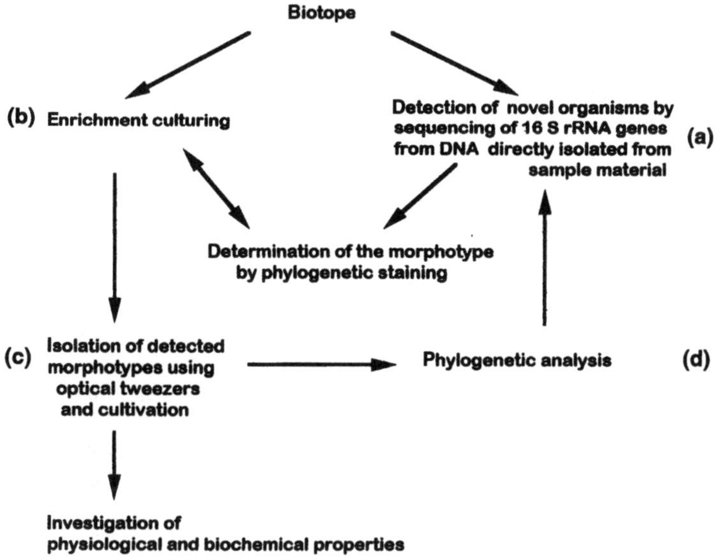

Figure 4. Procedure to specifically isolate organisms that harbor novel 16S rDNA sequences detected by direct phylogenetic analysis of environmental samples.

situ fluorescent probe hybridization (Burggraf et al., 1997a). The ability to grow these organisms in the laboratory in a pure culture to study their biochemical and physiological properties in detail will perhaps provide crucial information for understanding the evolution of *Archaea* and *Eucarya*. So far, only the examination of living cultures can reveal the versatile biochemical properties of microbes, which make them essential components of the biosphere (Palleroni, 1994). Therefore, it will be exciting to see if novel isolation tools like "optical tweezers" will help to get a pure culture of the *Korarchaeota*.

ACKNOWLEDGMENTS. We gratefully acknowledge the support of N. R. Pace, S. M. Barns, and K. O. Stetter. Furthermore, we are indebted to the U.S. Department of the Interior, National Park Service, for a sampling permit and B. Lindstrom for his field assistance. The excellent technical assistance of M. Bock, G. Maier, M. Weiss, G. Held, and P. Hummel is appreciated. We also thank G. Rieger for electron microscopy and N. Eis for comments on the manuscript. This work was supported by the Deutsche Forschungsgemeinschaft (Ste 297/10-1).

REFERENCES

Amann, R., Springer, N., Ludwig, W., Görtz, H.-D., and Schleifer, K.-H. 1991. Identification *in situ* and phylogeny of uncultured bacterial endosymbiont. *Nature* **351**:161–164.

Amann, R. I., Binder, B. J., Olson, R. J., Chisholm, S. W., Devereux, R., and Stahl, D. A. 1990a. Combination of 16S rRNA-targeted oligonucleotide probes with flow cytometry for analyzing mixed microbial populations. *Appl. Env. Microbiol.* **56**:1919–1925.

Amann, R. I., Krumholz, L., and Stahl, D. A. 1990b. Fluorescent-oligonucleotide probing of whole cells for determinative, phylogenetic, and environmental studies in microbiology. *J. Bacteriol.* **172**:762–770.

Ashkin, A., and Dziedzic, J. M. 1987. Optical trapping and manipulation of viruses and bacteria. *Science* **235:**1517–1520.

Ashkin, A., Dziedzic, J. M., and Yamane, T. 1987. Optical trapping and manipulation of single cells using infrared laser beams. *Nature* **330:**769–771.

Balch, W. E., Fox, G. E., Magrum, L. J., Woese, C. R., and Wolfe, R. S. 1979. Methanogens: Reevaluation of a unique biological group. *Microbiol. Rev.* **43:**260–296.

Balch, W. E., and Wolfe, R. S. 1976. New approach to the cultivation of methanogenic bacteria: 2-Mercaptoethane sulfonic acid (HS-CoM)-dependent growth of *Methanobacterium ruminantium* in a pressurized atmosphere. *Appl. Environ. Microbiol.* **32:**781–791.

Barns, S. M., Delwiche, C. F., Palmers, J. D., and Pace, N. R. 1996. Perspectives on archaeal diversity, thermophily and monophyly from environmental rRNA sequences. *Proc. Natl. Acad. Sci. USA* **93:**9188–9193.

Barns, S. M., Fundyga, R. E., Jeffries, M. W., and Pace, N. R. 1994a. Remarkable archaeal diversity detected in a Yellowstone National Park hot spring environment. *Proc. Natl. Acad. Sci. USA* **91:**1609–1613.

Barns, S. M., Fundyga, R. E., Jeffries, M. W., and Pace, N. R. 1994b. Remarkable archaeal diversity in a Yellowstone National Park hot spring. Abstracts of the *94th General Meeting of the American Society for Microbiology* 1994:253.

Blöchl, E., Burggraf, S., Fiala, G., Lauerer, G., Huber, G., Huber, R., Rachel, R., Segerer, A., Stetter, K. O., and Völkl, P. 1995. Isolation, taxonomy and phylogeny of hyperthermophilic microorganisms. *World J. Microbiol. Biotech.* **11:**9–16.

Brosius, J., Dull, T. J., Sleeter, D. D., and Noller, H. F. 1981. Gene organization and primary structure of a ribosomal RNA operon from *Escherichia coli. J. Mol. Biol.* **148:**107–127.

Burggraf, S., Heyder, P., and Eis, N. 1997a. A pivotal *Archaea* group. *Nature* **385:**780.

Burggraf, S., Huber, H., and Stetter, K. O. 1997b. Reclassification of the crenarchaeal orders and families in accordance with 16S rRNA sequence data. *Int. J. Syst. Bacteriol.* **47:**657–660.

Burggraf, S., Mayer, T., Amann, R., Schadhauser, S., Woese, C. R., and Stetter, K. O. 1994. Identifying members of the domain archaea with rRNA-targeted oligonucleotide probes. *Appl. Environ. Microbiol.* **60:**3112–3119.

DeLong, E. F. 1992. Novel archaea in coastal marine environments. *Proc. Natl. Acad. Sci. USA* **89:**5685–5689.

DeLong, E. F., Wickham, G. S., and Pace, N. R. 1989. Phylogenetic stains: Ribosomal RNA-based probes for the identification of single cells. *Science* **243:**1360–1363.

DeLong, E. F., Wu, K. Y., Prezelin, B. B., and Jovine, V. M. 1994. High abundance of *Archaea* in antarctic marine picoplankton. *Nature* **371:**695–697.

Fuhrman, J. A., McCallum, K., and Davis, A. A. 1992. Novel major archaebacterial group from marine plankton. *Nature* **356:**148–149.

Giovannoni, S. J., Britschgi, T. B., Moyer, C. L., and Field, K. G. 1990. Genetic diversity in Sargasso Sea bacterioplankton. *Nature* **345:**60–62.

Huber, R., Burggraf, S., Mayer, T., Barns, S. M., Rossnagel, P., and Stetter, K. O. 1995. Isolation of a hyperthermophilic archaeum predicted by *in situ* RNA analysis. *Nature* **376:**57–58.

Huber, R., Dyba, D., Huber, H., Burggraf, S., and Rachel, R. 1998. Sulfur-inhibited *Thermosphaera aggregans* sp. nov., a new genus of hyperthermophilic archaea isolated after its prediction from environmentally derived 16S rRNA sequences. *Int. J. Syst. Bacteriol.* **48:**31–38.

Huber, R., Kurr, M., Jannasch, H. W., and Stetter, K. O. 1989. A novel group of abyssal methanogenic archaebacteria (*Methanopyrus*) growing at 110°C. *Nature* **342:**833–834.

Huber, R., and Stetter, K. O. 1992. The order *Thermoproteales.* In Balows, A., Trüper, H. G., Dworkin, M., Harder, W., and Schleifer, K.-H. (eds.), *The procaryotes* (pp. 677–683). New York, Berlin, Heidelberg: Springer-Verlag.

Hugenholtz, P., Pitulle, C., Hershberger, K. L., and Pace, N. R. 1998. Novel division level bacterial diversity in a Yellowstone hot spring. *J. Bacteriol.* **180:**366–376.

Kane, M. D., Poulsen, L. K., and Stahl, D. A. 1993. Monitoring the enrichment and isolation of sulfate-reducing bacteria by using oligonucleotide hybridization probes designed from environmentally derived 16S rRNA sequences. *Appl. Env. Microbiol.* **59:**682–686.

Koch, R. 1881. Zur Untersuchung von pathogenen Organismen. *Mitt. Kaiserl. Gesundheitsamt Berlin* **1:**1–48.

Maidak, B. L., Olsen, G. J., Larsen, N., Overbeek, R., McCaughey, M. J., and Woese, C. R. 1996. The Ribosomal Database Project (RDP). *Nucleic Acids Res.* **24:**82–85.

Palleroni, N. J. 1994. Some reflections on bacterial diversity. ASM News **60:**537–540.

Stahl, D. A., and Amann, R. L. 1991. Development and application of nucleic acid probes in bacterial systematics.

In Stackebrandt, E., and Goodfellow, M. (eds.) *Sequencing and hybridization techniques in bacterial systematics* (pp. 205–248). Chichester, England: Wiley.

Stetter, K. O. 1982. Ultrathin mycelia-forming organisms from submarine volcanic areas having an optimum growth temperature of 105°C. *Nature* **300**:258–260.

Stetter, K. O. 1992. Life at the upper temperature border. In *Colloque Interdisplinaire du Comité National de la Recherche Scientifique, Frontiers of Life*. Gif-sur-Yvette, France: Editions Fronières.

Stetter, K. O., Lauerer, G., Thomm, M., and Neuner, A. 1987. Isolation of extremely thermophilic sulfate reducers: Evidence for a novel branch of archaebacteria. *Science* **236**:822–824.

Stetter, K. O., Thomm, M., Winter, J., Wildgruber, G., Huber, H., Zillig, W., Janekovic, D., König, H., Palm, P., and Wunderl, S. 1981. *Methanothermus fervidus*, sp. nov., a novel extremely thermophilic methanogen isolated from an Icelandic hot spring. *Zbl. Bakt. Hyg., I. Abt. Orig. C* **2**:166–178.

Ward, D. M., Weller, R., and Bateson, M. M. 1990. 16S rRNA sequences reveal numerous uncultured inhabitants in a natural community. *Nature* **345**:63–65.

9

Thermophilic Anoxygenic Phototrophs

Diversity and Ecology

Michael T. Madigan

---

1. INTRODUCTION

Thermophilic anoxygenic phototrophic bacteria anoxyphototrophs inhabit hot springs worldwide. The filamentous green bacterium *Chloroflexus aurantiacus* was the first thermophilic anoxyphototroph to be discovered. However, since the isolation of this organism nearly a quarter century ago (Pierson and Castenholz, 1971), at least seven other thermophilic anoxyphototrophs from various hot spring habitats have been isolated and characterized. In this chapter I describe the basic properties of these organisms, including their phylogenetic status and physiology, and then consider two examples of the way the physiological ecology of these organisms has been studied *in situ*. We begin with an overview of the organisms themselves.

2. DIVERSITY AND PHYLOGENY OF HOT SPRING ANOXYPHOTOTROPHS

Table 1 lists thermophilic anoxyphototrophs that have been studied in laboratory culture. These include several new species isolated and characterized in the author's laboratory within the past 15 years. Figure 1 shows a phylogenetic tree of the domain *Bacteria* based on 16S ribosomal RNA sequence comparisons (Woese, 1987) and indicates which of the lineages contain phototrophic species. Of the twelve major lineages currently known in the domain *Bacteria*, five contain phototrophic representatives, and four of these contain

Michael T. Madigan • Department of Microbiology, Southern Illinois University, Carbondale, Illinois 62901.

Thermophiles: Biodiversity, Ecology, and Evolution, edited by Reysenbach *et al.* Kluwer Academic/Plenum Publishers, New York, 2001.

Table 1
Hot Spring Anoxygenic Phototrophs

Group	Bacteriochlorophyll	Color of mass suspension	Temp. optimum (°C)	Temp. maximum (°C)	Habitat(s)	Reference
Purple sulfur bacteria						
Chromatium tepidum[a]	*a*	Red	47–49	57	Yellowstone, New Mexico	Madigan, 1984, 1986
Chromatium sp.[b]	*a*	Red/brown	46	51	New Zealand	Madigan, unpublished
Purple nonsulfur bacteria						
Rhodopseudomonas sp. strain GI	*b*	Green	42	47	New Mexico, Yellowstone	Resnick and Madigan, 1989
Rhodopseudomonas cryptolactis	*a*	Red	40	46	Thermophilis (WY)	Stadtwald-Demchick et al., 1990
Rhodospirillum centenum	*a*	Red	40	47	Thermophilis (WY)	Favinger et al., 1989
Green sulfur bacteria						
Chlorobium tepidum	*c*	Green	48	52	New Zealand	Castenholz et al., 1990; Wahlund et al., 1991
Green filamentous bacteria						
Chloroflexus aurantiacus	c_s	Orange/green	50–55	65–70	Yellowstone, Oregon, Iceland, New Zealand	Pierson and Castenholz, 1974a,b
Heliothrix oregonensis	*a*	Orange	45–50	57	Oregon, Yellowstone	Pierson et al., 1985
Heliobacteria						
Heliobacterium modesticaldum	*g*	Green/brown	50–52	56	Yellowstone, Iceland, Oregon	Kimble et al., 1995

[a]This organism has been reclassified as its own genus, *Thermochromatium* (*Thermochromatium tepidum*) (Imhoff et al., 1998).
[b]This organism lacks the B920 light-harvesting complex present in *Chromatium tepidum* (Garcia et al., 1986).

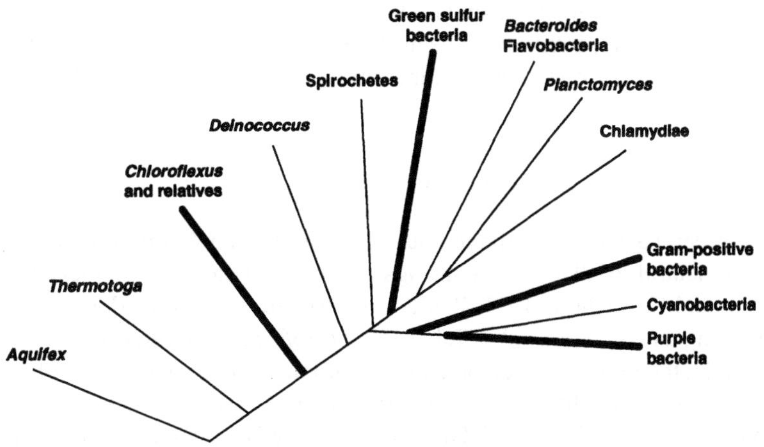

Figure 1. Phylogenetic tree of the domain Bacteria based on 16S ribosomal RNA sequence comparisons (Woese, 1987). Lineages that contain anoxygenic phototrophic bacteria are shown by bold lines. Each such lineage also contains thermophilic representatives, for example, purple bacteria (*Chromatium tepidum*), gram-positive bacteria (*Heliobacterium modesticaldum*), green sulfur bacteria (*Chlorobium tepidum*), *Chloroflexus* and relatives (*Chloroflexus aurantiacus*).

*anoxy*phototrophs (Fig. 1). Each lineage of *Bacteria* that contains anoxyphototrophs also contains thermophilic anoxyphototrophs.

2.1. Purple Bacteria

Within the large phylogenetic group called the "purple bacteria" (or "proteobacteria," as they are referred to in phylogenetic parlance) (Fig. 1), the purple sulfur bacterium *Chromatium tepidum* is the most thermophilic phototrophic species; *C. tepidum* inhabits neutral hot springs that contain moderate levels (< 0.05 mM) of sulfide at temperatures up to 60°C (Madigan, 1984, 1986). *Chromatium* species have also been observed in alkaline, low sulfide hot springs (unpublished results and Castenholz, 1977; Castenholz and Pierson, 1995), and isolates have been obtained in pure culture by the author from hot springs in New Zealand and New Mexico (U.S.) (Table 1). Whether these are all the same species is not known. However, like the Yellowstone (Mammoth Hot Springs) strain of *C. tepidum*, the New Mexico isolate contains a B920 light-harvesting pigment-protein complex that absorbs in the infrared near 920 nm (Garcia et al., 1986; Nozawa et al., 1986), whereas the New Zealand strain lacks this component; the latter isolate contains instead a B890 light-harvesting complex (as does the mesophilic *Chromatium* species, *C. vinosum*; Garcia et al., 1986), and cannot grow above 50°C (unpublished results). Thus, it is possible that various *Chromatium* species inhabit hot springs, each of which may have its own distinct photosynthetic properties and temperature requirements.

Purple *nonsulfur* bacteria are apparently widespread in warm thermal springs (Castenholz and Pierson, 1995) but to date only three isolates, a bacteriochlorophyll *b*-containing organism resembling *Rhodopseudomonas viridis* (Resnick and Madigan, 1989), a *Rhodopseudomonas palustris*-like organism described as a new species, *Rhodopseudomonas cryptolactis* (Stadtwald-Demchick et al., 1990), and the highly motile *Rhodospirillum centenum* (Favinger et al., 1989), have been studied in pure culture; all of these organisms are mildly thermophilic, growth temperature optima are between 40 and 45°C, and maxima are below 50°C.

2.2. Green Bacteria and Heliobacteria

Among green bacteria, the filamentous green bacterium *Chloroflexus aurantiacus* is widespread in neutral to alkaline hot springs worldwide (Bauld and Brock, 1973; Brock, 1978; Pierson and Castenholz, 1971, 1974a,b, 1995) and is the most thermotolerant of all known hot spring anoxyphototrophs (Table 1). *Chloroflexus* is of interest for many reasons (see for example, the review in Pierson and Castenholz, 1995), not the least of which is the fact that it is phylogenetically the most ancient (branches nearest to the root) of all known phototrophic organisms (Fig. 1 and Gibson et al., 1985). A phylogenetic relative of *C. aurantiacus* is *Heliothrix oregonensis* (Pierson et al., 1985). This filamentous bacterium contains only bacteriochlorophyll *a* and lacks the bacteriochlorophyll *c* and chlorosomes (see later) of *Chloroflexus*.

The thermophilic species *Chlorobium tepidum* (Table 1) is a member of the phylogenetically distinct green sulfur bacteria (Fig. 1) and is very geographically restricted; isolates of *C. tepidum* have been obtained only from high (> 0.3 mM) sulfide acidic New Zealand hot springs at temperatures from 40–60°C (Castenholz et al., 1990; Wahlund et al., 1991). Ecological studies of natural populations of *C. tepidum* (Castenholz et al., 1990) suggest that this organism is well suited to its habitat because the combination of high sulfide and low pH in which it thrives virtually eliminates competition from other phototrophs like cyanobacteria and purple bacteria. What appear to be organisms similar to *Chlorobium tepidum* have been observed in Wilbur Springs, California (R.W. Castenholz, personal communication), but samples failed to yield laboratory cultures in the Madigan lab.

Heliobacterium modesticaldum is the most recently characterized thermophilic anoxyphototroph among those listed in Table 1 and is probably widely distributed in neutral to alkaline hot spring microbial mats at temperatures to ~60°C (Kimble et al., 1995; Stevenson et al., 1997). Like its mesophilic relatives, *H. modesticaldum* belongs to the gram-positive lineage of the domain Bacteria (Fig. 1) and groups closely with species of *Clostridium* and *Desulfotomaculum* (Kimble et al., 1995; Redburn and Patel, 1993). In this connection, the ability of *H. modesticaldum* and other heliobacteria to form endospores (Kimble et al., 1995), a property unique to heliobacteria among phototrophic microorganisms (Ormerod et al., 1990, 1996), may have significance for the survival and distribution of this organism in thermal environments. Unlike *C. aurantiacus*, *Chromatium tepidum*, and *Chlorobium tepidum*, *H. modesticaldum* does not form mass accumulations (blooms) in thermal springs, and nothing is yet known of its ecological impact on hot spring microbial mats.

2.3. Ultrastructure of Some Thermophilic Anoxyphototrophs

Figure 2 shows electron micrographs of cells of representative thermophilic anoxyphototrophs. The ultrastructure of each organism is distinctive; the arrangement of membranes and photosynthetic complexes observed is characteristic of a particular taxonomic group of anoxyphototrophs (Table 1). For example, in phototrophic purple bacteria, intracytoplasmic membranes of various morphologies are observed (Kondratieva et al., 1992). In *Chromatium tepidum* (Fig. 2a), these membranes take the form of membrane vesicles; in *Rhodopseudomonas sp.* GI (Fig. 2b), lamellar membranes typical of mesophilic budding purple bacteria are observed.

Both the thermophilic green bacteria *Chlorobium tepidum* and *Chloroflexus aurantiacus* synthesize unique photosynthetic structures called chlorosomes (Fig. 2c,d,e). Chlorosomes contain the light-harvesting bacteriochlorophylls (bacteriochlorophylls *c*, *d*, or *e*) in green bacteria and funnel energy to the reaction center, which contains bacteriochlorophyll *a*, that is located in the cytoplasmic membrane (Oelze and Golecki, 1995). Interestingly, despite their phylogenetic divergence (Gibson et al., 1985), the basic structure of chlorosomes from *C. tepidum* and *C. aurantiacus* are quite similar, suggesting that lateral gene transfer may have occurred between these species in the distant past. Moreover, *Chlorobium tepidum* and *C. aurantiacus* each contain several unique carotenoids, including structurally similar carotenoid glucoside esters (Taikaichi et al., 1995, 1997b). Is it possible that these carotenoids are required for photosynthetic processes at high temperature in green bacteria?

In contrast to *C. tepidum* and *C. aurantiacus*, *H. modesticaldum* lacks internal photosynthetic membranes and chlorosomes (Fig. 2f) as is also typical of mesophilic heliobacteria; unlike all other anoxyphototrophs, the photopigments of heliobacteria are housed solely in the cytoplasmic membrane (Amesz, 1995; Miller et al., 1986). In addition, the key carotenoid of *H. modesticaldum*, diaponeurosporene, is the same as the major carotenoid of various mesophilic species of heliobacteria (Taikaichi et al., 1997a).

From this brief survey, it can be concluded that considerable phylogenetic and structural diversity exists among hot spring anoxyphototrophs but that the basic photosynthetic structures of mesophilic anoxyphototrophs can survive, presumably in modified forms, at high temperature. Moreover, the discovery and characterization of thermophilic anoxygenic phototrophs has clearly broadened our understanding of the diversity of photosynthetic microorganisms, and the organisms themselves are undoubtedly reservoirs of heat stable photosynthetic membranes and enzymes; some examples of the latter are discussed in the next section.

3. PHYSIOLOGY OF HOT SPRING ANOXYPHOTOTROPHS

This section highlights some of the physiological features of thermophilic anoxyphototrophs and focuses on results obtained in the author's laboratory with the species *Chromatium tepidum*, *Chlorobium tepidum*, *H. modesticaldum*, and *Rhodopseudomonas sp.* GI.

3.1. Temperature Relationships

Hot spring anoxyphototrophs can be considered only "mild" thermophiles. Except for *Chloroflexus*, which grows optimally at around 55°C (Pierson and Castenholz, 1974b), most hot spring anoxyphototrophs grow optimally at around 50°C; purple nonsulfur

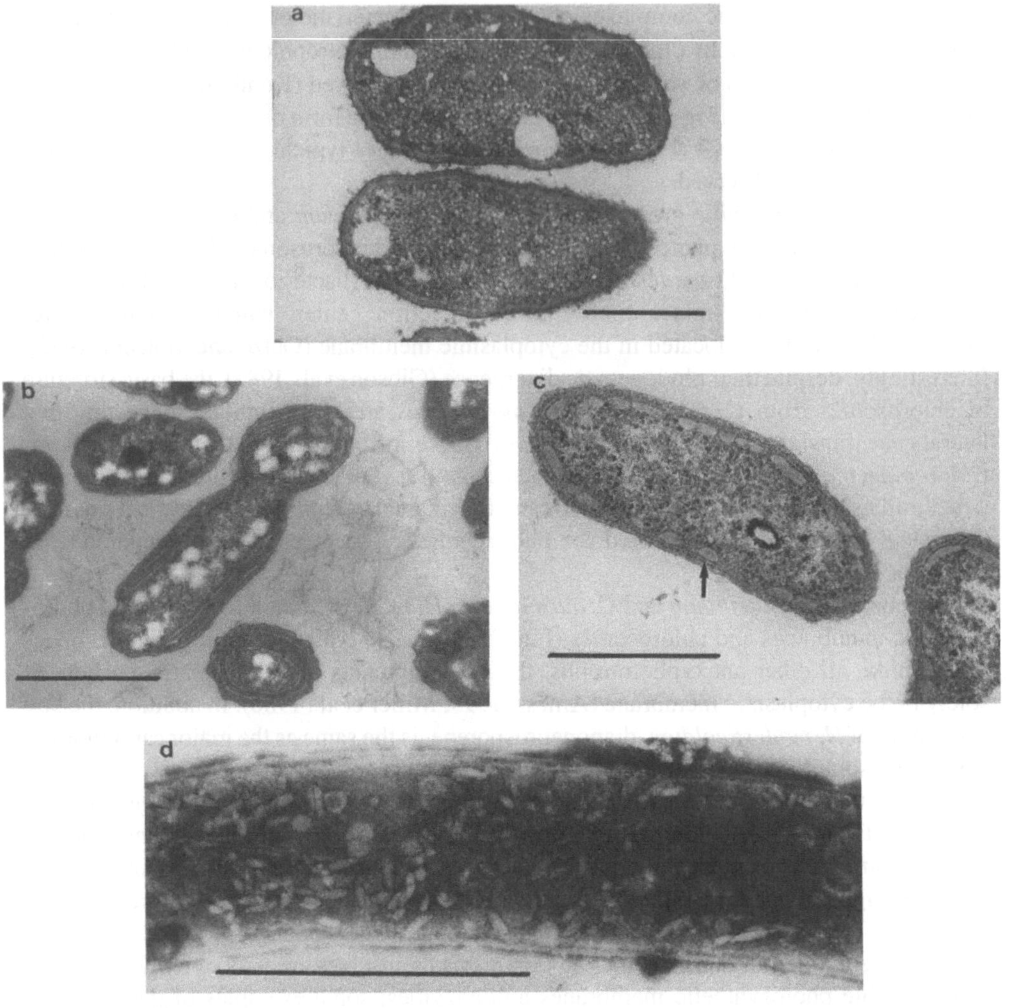

Figure 2. Transmission electron micrographs of cells of representative thermophilic anoxyphototrophs (marker bars in all micrographs = 1 μm). (a) Thin section of cells of *Chromatium tepidum* strain MC; note vesicular intracytoplasmic membranes. (b) Thin section of cells of *Rhodopseudomonas* sp. strain GI; note lamellar-type intracytoplasmic membranes. (c) Thin section of a cell of *Chlorobium tepidum* strain TLS; note chlorosomes (arrow). (d) Negatively stained cells of two Yellowstone strains of *Chloroflexus aurantiacus*, strain 254-2 [d, isolated from Grassland Spring (61°C) located about 0.5 km east of Octopus Spring], and strain 396-1 [e, isolated from Conophyton Pool (30–40°C) located at the south end of Fairy Creek meadow]; note chlorosomes in both strains. (e) Thin section of a cell of *Heliobacterium modesticaldum* strain Ice 1.

bacteria are somewhat less thermophilic (Fig. 3 and Table 1). Nevertheless, the cardinal temperature of each species of thermophilic anoxyphototroph is distinct from that of its closest mesophilic relative (Kimble et al., 1995; Madigan, 1984, 1986; Resnick and Madigan, 1989; Wahlund et al., 1991). The maximum growth temperature for cultures of most species lies between 52° and 57°C (Fig. 3). The species *Chromatium tepidum*, *Chlorobium*

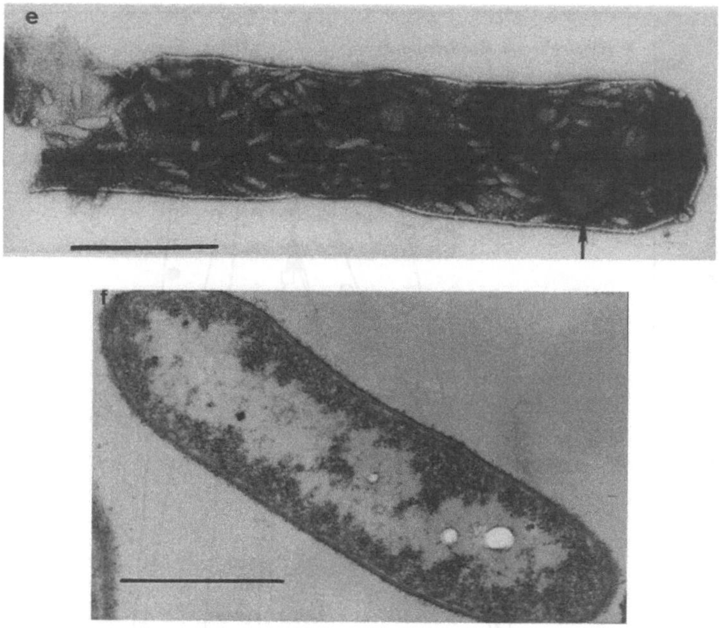

Figure 2. (*Cont.*)

tepidum, and *Heliobacterium modesticaldum* all grow faster in pure culture than their mesophilic relatives, and this is especially true of *Chlorobium tepidum*; the latter organism grows with a ~2-h generation time at 48–49°C, which is much faster than that of any mesophilic green bacterium (Heda and Madigan, 1986b; Wahlund et al., 1991). Because of this and because it is easy to culture (*C. tepidum* uses thiosulfate as a photosynthetic electron donor; this simplifies the preparation of culture media and allows for growth of laboratory cultures to high cell densities; see Wahlund and Madigan, 1993), *Chlorobium tepidum* has drawn considerable attention from scientists interested in the basic aspects of photosynthesis (Frigaard et al., 1997; Taikaichi et al., 1997; and see the list of references using *Chlorobium tepidum* cited in Blankenship et al., 1995).

3.2. Autotrophy

Both *Chromatium tepidum* and *Chlorobium tepidum* oxidize H_2S or S^0 to SO_4^{2-} as electron donors for photosynthetic CO_2 fixation (Madigan, 1986; Wahlund et al., 1991); however, only *Chlorobium tepidum* can oxidize $S_2O_3^{2-}$ to SO_4^{2-} (Wahlund et al., 1991). Thus, the metabolism of both species is photoautotrophic, although organic compounds such as acetate stimulate growth, especially in cultures of *Chromatium tepidum* (Madigan, 1986). Apparently, *H. modesticaldum* is incapable of autotrophic growth (Kimble et al., 1995), and grows photosynthetically only when supplied with an organic compound as a carbon source (that is, photoheterotrophic growth). *Chromatium aurantiacus* grows slowly as a photoautotroph using either H_2S (Madigan and Brock, 1975) or H_2 (Sirevåg and Castenholz, 1979) as an electron donor. In addition to a phototrophic lifestyle, *C. auran-*

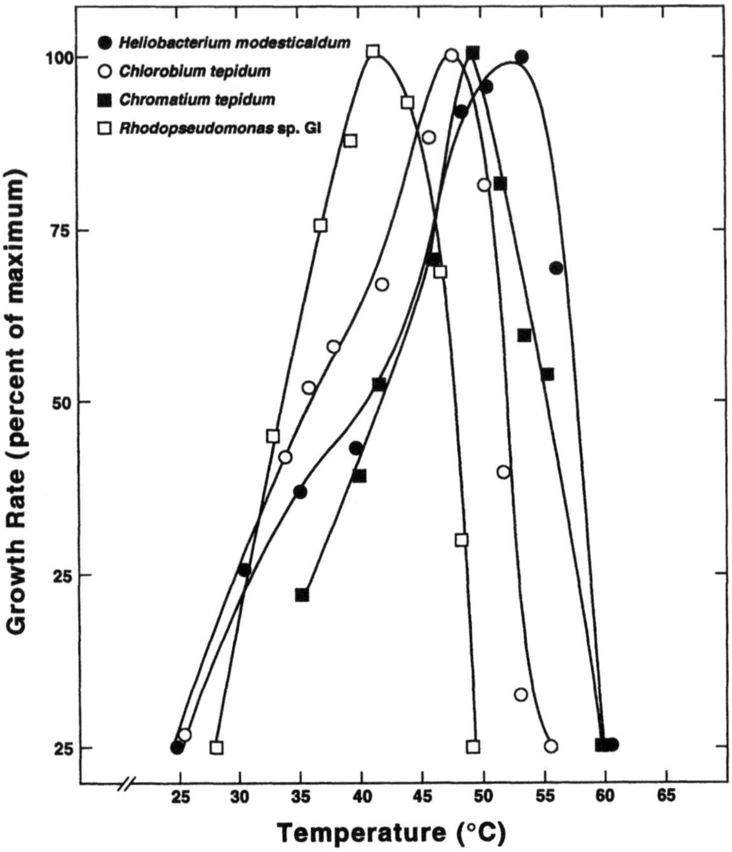

Figure 3. Growth rate as a function of temperature for a number of thermophilic anoxyphototrophs.

tiacus and *H. modesticaldum* can also grow as chemotrophs in darkness; under dark conditions, *Chloroflexus* grows aerobically by respiring a wide variety of organic compounds (Madigan et al., 1974; Pierson and Castenholz, 1974a), and *H. modesticaldum* grows strictly anaerobically by pyruvate fermentation (Kimble et al., 1994, 1995). By contrast, *Chromatium tepidum* and *Chlorobium tepidum* are apparently obligate phototrophs (Madigan, 1986; Wahlund et al., 1991).

The only studies that have been undertaken of the biochemistry of autotrophy in thermophilic anoxyphototrophs are those of *C. aurantiacus* and *Chromatium tepidum*. In *Chloroflexus*, CO_2 is fixed by the novel autotrophic hydroxypropionate pathway (Holo and Sirevåg, 1986; Strauss and Fuchs, 1993). This pathway has not been found in any other phototrophic or chemolithotrophic autotroph, and this, coupled with the fact that *C. aurantiacus* is phylogenetically the most ancient of known anoxyphototrophs (Gibson et al., 1985; Woese, 1987; see also Fig. 1), suggests that the hydroxypropionate pathway may have been among the first autotrophic pathways that evolved in *Bacteria*. In *Chromatium tepidum*, the Calvin cycle is operative for CO_2 fixation, and the purified enzyme ribulose

1,5-bisphosphate carboxylase (RubisCO) was shown to have the same molecular structure (that is, 8 large and 8 small subunits) as RubisCOs from green plants, except for its thermophilic and thermostable properties (Heda and Madigan, 1985, 1989). The molecular basis for the heat stability of this enzyme is unknown.

3.3. Nitrogen Fixation

The author's laboratory has researched the process of nitrogen fixation ($N_2 + 8H \rightarrow 2NH_3 + H_2$) in anoxyphototrophs for some time (see Madigan, 1995 for a review in this area) and has documented the ability of certain thermophilic anoxyphototrophs to fix N_2. In *C. aurantiacus*, the most phylogenetically ancient of all anoxyphototrophs (Gibson et al., 1985; Woese, 1987; and see Fig. 1), neither phenotypic (growth on N_2 and acetylene reduction) nor genotypic (Southern hybridization of genomic DNA to cloned *nifHDK* genes) evidence for the capacity to fix nitrogen has been obtained (Table 2 and Heda and Madigan, 1986a). [The reduction of acetylene (C_2H_2) to ethylene (C_2H_4) is used as an assay for N_2 fixation (Postgate, 1982) and *nifHDK* are the structural genes for nitrogenase, the enzyme that catalyzes N_2 fixation]. Growth of *Chromatium tepidum* on N_2 has not yet been achieved, and acetylene reduction assays of cells cultured on growth limiting levels of ammonia (a means for strongly derepressing nitrogenase synthesis in nitrogen-fixing bacteria; see Postgate, 1982) have been negative (Table 2 and unpublished results). However, molecular studies using a cloned *nifHDK* probe clearly showed that *C. tepidum* DNA contains *nifHDK* (or *nifHDK*-like) genes (Paul W. Ludden, Gary P. Roberts, and Michael T. Madigan, unpublished results). Thus, it is possible that *C. tepidum* can fix N_2 but that the proper conditions for expressing its nitrogenase system have not yet been achieved. Alternatively, *C. tepidum* may be missing (through gene deletion, for example) one or more key *nif* genes necessary to make a functional nitrogenase complex. In addition to work with pure cultures, field experiments (Madigan, unpublished results) gave no evidence of C_2H_2 reduction by natural populations of *Chromatium tepidum*, including assays using cells from the very spring that yielded the Yellowstone strain. The latter contains very low levels of NH_4^+ (Castenholz, 1969), as do most Yellowstone hot spring waters (Brock, 1978), and thus one would predict that if diazotrophy were possible in *C. tepidum*, it would occur under these conditions. More study of this problem is desirable and might reveal special physiochemical conditions necessary for N_2 fixation by *C. tepidum* (or other thermophiles, for that matter, that may be cryptic nitrogen-fixing bacteria as we know them from laboratory culture).

In contrast with *Chromatium tepidum*, cultures of *Chlorobium tepidum* and *H. modesticaldum* grow well at 50°C on N_2 as the sole source of nitrogen (Kimble et al., 1995; Wahlund et al., 1991; Wahlund and Madigan, 1993), and cultures of both organisms readily reduce C_2H_2 to C_2H_4 (Table 2); thus, both species are clearly diazotrophic. These results are of interest for at least two reasons. First, thermophilic nitrogen-fixing bacteria are apparently quite rare (Postgate, 1982); this makes organisms like *Chlorobium tepidum* and *H. modesticaldum* that can grow well and fix dinitrogen at 50°C good candidates for the study of N_2 fixation at high temperatures. But in addition to this, *in situ* N_2 fixation by *C. tepidum* and *H. modesticaldum* may be an important ecological strategy for their survival in

Table 2
Nitrogen Fixation by Pure Cultures of Thermophilic Anoxyphototrophs

Organism	Assay temperature (°C)	Nitrogenase[a] activity	Reference
Chlorobium tepidum			
strain TLS	48	6.3	Wahlund and Madigan, 1993
	51	5.3	
	55	0.53	
	60	0.19	
	65	0.19	
Heliobacterium modesticaldum[b]			
strain Ice1	45	3.8	Kimble et al., 1995
	50	3.8	
	55	0.53	
	60	0	
strain YS5	45	2.2	Kimble et al., 1995
	50	1.4	
	55	<0.001	
Rhodopseudomonas sp.[c]			
strain GI	42	0.26	Resnick and Madigan, 1989
	48	0.095	
	50	<0.001	
Chromatium tepidum[d]			
strain MC	40	<0.001	Madigan, unpublished results
	45	<0.001	
	50	<0.001	
	55	<0.001	
Chloroflexus aurantiacus[e]			
strains J-10-fl, OK-70-fl, Y-400-fl, 396-1	45	<0.001	Heda and Madigan, 1986a
	50	<0.001	
	55	<0.001	

[a]Nitrogenase activity measured in anoxic cell suspensions of N_2-grown cultures or cultures grown on limiting ammonia (see Heda and Madigan, 1986a; Kimble et al., 1995; Resnick and Madigan, 1989; and Wahlund and Madigan, 1993, for precise growth methods) supplemented with 10% C_2H_2 and incubated in the light at the temperature listed. Nitrogenase activity expressed as μmol C_2H_4 produced $\cdot h^{-1} \cdot$ mg cell dry weight^{-1}.
[b]Strain Ice1 is from an Icelandic hot spring; strain YS5 is from an unnamed Yellowstone spring (Kimble et al., 1995).
[c]Isolated from an alkaline hot spring (45°C) in New Mexico (Resnick and Madigan, 1989).
[d]Cells grown at 48°C on limiting levels of ammonia until the ammonia was exhausted.
[e]Cells grown at 50°C (except for strain 396-1, which was grown at 39°C) as in d. Results were the same (that is, negative) in acetylene reduction tests with all four strains. Strain J-10-fl is from a Japanese hot spring, OK-70-fl from an Oregon hot spring, and strains Y-400-fl and 396-1 from Octopus Spring and Conophyton Pool, respectively (both Yellowstone National Park).

nitrogen-poor hot springs (Brock, 1978; Castenholz, 1969). In this connection it would be interesting to know to what extent nitrogen fixation measured by acetylene reduction in several Yellowstone microbial mats and ascribed to the activities of cyanobacteria (Stewart, 1970), was really due to nitrogen fixation by *H. modesticaldum* or *Chlorobium tepidum*.

In addition to *Chlorobium tepidum* and *H. modesticaldum*, all thermophilic nonsulfur purple bacteria tested are also diazotrophic; in general they fix N_2 up to or slightly above their maximum growth temperature (Table 2 and Favinger et al., 1989; Resnick and Madigan, 1989; Stadtwald-Demchick et al., 1990). The ecological significance of nitrogen fixation by these phototrophs is unknown.

3.4. Thermostable Enzymes

It is likely that most enzymes from thermophilic anoxyphototrophs are at least somewhat thermostable and some may even be extremely so. Thus far, however, except for *Chromatium tepidum* RubisCO (see earlier discussion), which is stable to 60°C (16), and the citric acid cycle enzyme malate dehydrogenase, whose molecular structure and genes have been characterized from *C. aurantiacus* (Synstad et al., 1996) and from *Chlorobium tepidum* (Charnock et al., 1992), no studies of thermostable proteins from these organisms have been undertaken. Both the malate dehydrogenases of *C. aurantiacus* and *C. tepidum* are somewhat thermostable; the enzyme from *C. aurantiacus* is stable to about 60°C and the *C. tepidum* enzyme to 50–55°C (Synstad et al., 1996). As previously mentioned, the nitrogenases of *Chlorobium tepidum* and *H. modesticaldum* function *in vivo* above 50°C, indicating that these proteins are also likely to be thermostable *in vitro*. Thus, studies of the molecular structure of nitrogenases from these organisms could yield clues as to the structural requirements for a thermostable nitrogenase and perhaps also shed light on the reason that so few thermophilic microorganisms are diazotrophic (Postgate, 1982).

Because of their thermostability, future studies of enzymes and photosynthetic membranes from anoxyphototrophs are likely to focus increasingly on thermophilic species, especially since modern molecular biology allows for the facile cloning of genes and their expression in more genetically tractable organisms (for some examples of this that have been done using genes from *Chlorobium tepidum*, the reader is referred to the volume on anoxyphototrophs by Blankenship et al., 1995). Although biotechnological applications of such enzymes lie in the future, there is little doubt that thermostable proteins and photosynthetic complexes from anoxyphototrophs will eventually find useful applications.

4. ECOLOGICAL STUDIES OF THERMOPHILIC ANOXYPHOTOTROPHS

In situ ecological studies of thermophilic anoxygenic phototrophs have focused for the most part on habitats in which the organisms form mass accumulations, because ample cell material is available for field measurements. For an excellent overview of ecological studies of this type, the reader is referred to the recent review by Castenholz and Pierson (1995) and to the monograph by Brock (1978).

The author has performed two studies of natural populations of thermophilic anoxyphototrophs in Yellowstone; one involved *C. aurantiacus* (Madigan and Brock, 1977), and the other *Chromatium tepidum* (Madigan et al., 1989). The methods and results of both studies are described here as examples of two quite different experimental approaches to the study of ecological problems that face thermophilic anoxyphototrophs in nature.

4.1. Adaptation by *Chloroflexus* to Reduced Light Intensity

In alkaline hot springs of Yellowstone at temperatures from 45–70°C, thick microbial mats develop that contain cyanobacteria and *Chloroflexus* (Bauld and Brock, 1973; Brock, 1978; Castenholz, 1969; Castenholz and Pierson, 1995; Pierson and Castenholz, 1974a). At the altitude of Yellowstone (slightly higher than 2100 m), light intensities frequently reach maximal values of more than 8000 foot-candles (approx. 86,000 Lux) at midday; during the summer months, such values are often reached daily for several days or even several weeks at a time (Brock and Brock, 1969; Doemel and Brock, 1974). By contrast, substantial light reduction occurs during the winter months due to increased cloudiness and shortened daylight periods. In addition to these seasonal changes in light intensity, light gradients within the microbial mat are quite steep; measurements have shown that the top 1 mm or so of these mats absorbs or scatters nearly 95% of incident radiation (Doemel and Brock, 1974). This situation raises two interesting questions. First, are cells of *Chloroflexus* in the uppermost regions of these mats physiologically adapted to carry out photosynthesis optimally at high light intensities in summer months? And second, does cell growth within the mat (which leads to self-shading) or the onset of winter lighting conditions trigger physiological adaptations that allow photosynthesis to occur optimally at lower light intensities? An interesting related question but one not addressed in the study to be described is whether seasonal differences in light intensity catalyze changes in the microbial composition of the mats; in this connection, recent experiments using 16S rRNA sequencing (Ferris and Ward, 1997; Ward et al., this volume) have shown that seasonal changes in the microbial population in the Octopus Spring mat do occur.

4.1.1. Experimental Light Reduction Studies

To approach these questions experimentally, the light intensity was artificially reduced over *Synechococcus/Chloroflexus* mats in thermal springs in the White Creek area of the Lower Geyser Basin of Yellowstone (Brock, 1978) in the summer of 1975. The discussion here focuses on work done on Octopus Spring (Madigan and Brock, 1977). Using neutral density filters, light intensity was reduced by 73%, 93%, 98%, and 100% of full sunlight over the lush mat that exists at Octopus Spring, and an adjacent control site in the same mat was left uncovered for comparative measurements. After the filters were in place for about 2 weeks, changes in pigmentation in the surface layer of the mat were readily apparent; compared with the uncovered site, the top ~1 mm of the 73% and 93% reduction sites changed from an olive green to a distinct blue-green color, whereas the 98% reduction site turned bright orange, the color of cells of *Chloroflexus*. Surprisingly, however, these dramatic color changes did not result from significant changes in chlorophyll *a* (in *Synechococcus* cells) or bacteriochlorophyll *c* (in *Chloroflexus* cells); when normalized to cell protein, essentially no changes in chlorophyll or bacteriochlorophyll were observed at these light reduction sites except under the 98% filter, where chlorophyll *a* (and thus the *Synechococcus* population) was totally lost (Madigan and Brock, 1977). Microscopy of samples of the surface layer confirmed this; a population of *Synechococcus* and *Chloroflexus* remained at all light reduction sites except under the 98% filter, which essentially contained a pure stand of *Chloroflexus* (see Fig. 11 of Madigan and Brock, 1977). A completely darkened region of the mat showed the loss of both phototrophic components, indicating that at least some light was necessary to maintain a *Chloroflexus* population.

Unfortunately, the nature of the pigmentation changes observed in this study were not resolved. However, it was hypothesized that the intense blue-green color was due to increases in phycobilins (e.g., phycocyanin) in the cells of *Synechococcus*, and the bright orange was due to carotenoids of the cells of *Chloroflexus* (Madigan and Brock, 1977). In the latter connection, studies of the pigment composition of *C. aurantiacus* have shown that the carotenoid composition of cells grown in low light (ca. 50 foot-candles) differs from that of cells grown in high light (ca. 540 foot-candles). However, it is not known which specific pigment(s) of the many carotenoids known from *C. aurantiacus* (Halfen et al., 1972) vary with light intensity.

4.1.2. Photosynthesis by Low Light and Control Populations of *Chloroflexus*

Now that readily visible changes occurred in the pigmentation of phototrophic organisms in this top layer of the mat, did these changes, whatever their basis, have physiological relevance? Did the natural populations of *Synechococcus* and *Chloroflexus* that were exposed to reduced light intensities adapt to these reduced light intensities? To determine this, rates of $^{14}CO_2$ incorporation by natural populations of both phototrophs incubated at different light intensities were measured in short term (1-h) experiments; to focus on *Chloroflexus*, photoincorporation of CO_2 by *Synechococcus* was blocked by adding 3-(3,4-dichlorophenyl)-1,1-dimethylurea (DCMU), an inhibitor of oxygenic photosynthesis. The discussion here deals only with *Chloroflexus*, although the results with *Synechococcus* were similar (Madigan and Brock, 1977).

In samples of the top layer from control (unshaded) mats, a clear dose–response curve was observed in plots of photosynthesis versus light intensity; cells of *Chloroflexus* incorporated $^{14}CO_2$ at an increasing rate from near zero in darkness to a maximal rate at about 1500 foot-candles and then maintained this rate to 8000 foot-candles (Fig. 4). By contrast, *Chloroflexus* cells from this same upper layer but taken from under the 73% light reduction filter showed a distinct light intensity optima for $^{14}CO_2$ incorporation. In this population, rates of photosynthesis were maximal at about 1800 foot-candles and fell sharply at higher light intensities (Fig. 4). This trend was even more dramatic in measurements of photosynthetic CO_2 incorporation by the virtually pure population of *Chloroflexus* under the 98% reduction filter; photosynthesis by these cells was maximal at about 600 foot-candles and was near zero at all light intensities tested above 800 foot-candles (Fig. 4).

These data clearly demonstrate that natural populations of *Chloroflexus* can physiologically adapt to reduced light intensities but as a consequence of this, become high light inhibited. By contrast, control populations of *Chloroflexus* not subject to experimental light reduction photosynthesized equally well from relatively low (1500 foot-candles) to extremely high (8000 foot-candles) light intensities (Fig. 4). How can this be accounted for? One possibility is that even within the uppermost 1 mm of mat different populations of *Chloroflexus* exist with respect to light intensity—high light adapted versus low light adapted—depending on the exact position of a cell in the mat and the degree of self-shading it experiences. If this hypothesis is correct, then in the protocol employed where a section of the top region of the mat was homogenized to yield replicate samples that were placed in screw-cap vials for the radioisotope experiments (Madigan and Brock, 1977), only a certain fraction of the cells in each vial would have been photosynthetically active at a particular light intensity. Alternatively, because cells of *Chloroflexus* are capable of gliding motility, cells in the upper mat layer may be a relatively homogeneous population with respect to the light intensity they experience. Then, one could hypothesize that cells are both light saturated for photosynthesis at relatively low intensities but remain immune to the damaging effects of high light intensities due to the presence of protective pigments. In this regard, Pierson and Castenholz (1974b) observed that at least one carotenoid was constitutively expressed in cells of *C. aurantiacus*, including cells grown aerobically in darkness or

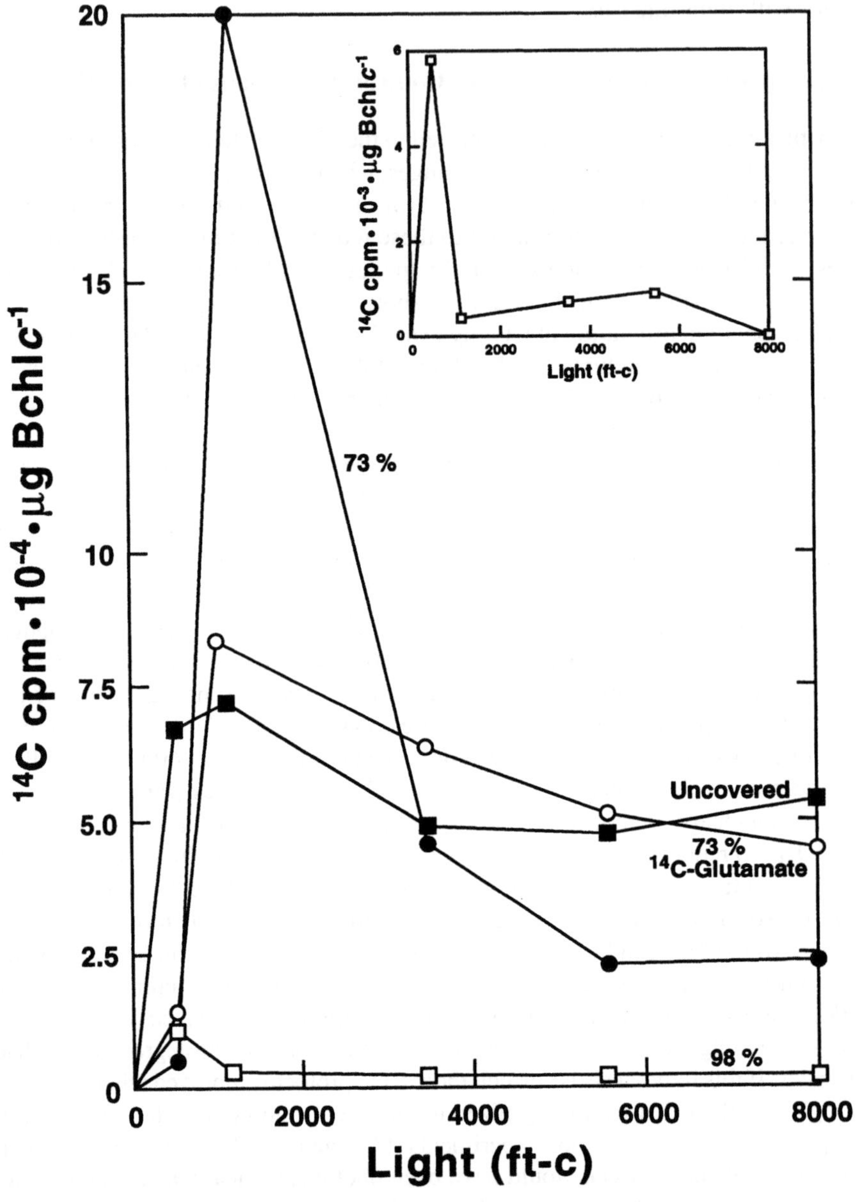

Figure 4. Light-dependent $^{14}CO_2$ incorporation as a function of light intensity by cells of *Chloroflexus aurantiacus* in the top ~1 mm of microbial mat from Octopus Spring (see Madigan and Brock, 1977, for methods). Data are normalized to bacteriochlorophyll *c* content. "Uncovered" refers to cells from the uncovered control site; "73%," cells from under the 73% light reduction filter assayed for either $^{14}CO_2$ or ^{14}C-glutamate incorporation as indicated; 98%, cells from under the 98% light reduction filter (data replotted in inset). Incident light intensity during the course of this experiment (July 23, 1975) remained constant at 8400 ft-candles (90,384 lux) (data from Madigan and Brock, 1977).

phototrophically (anaerobically) at any light intensity; it is possible that this pigment plays a key role in protecting against photooxidative killing.

4.1.3. Minimal Light Intensities for Growth

The light reduction studies done on the Octopus Spring mat also explain why *Synechococcus* is absent from regions below the top layer of the microbial mat. To maintain itself in the mat, *Synechococcus* apparently must experience light intensities of at least 2% of incident radiation (the precise value is less than 7% but more than 2% of summer intensities; see Madigan and Brock, 1977). At intensities less than this, only *Chloroflexus* can survive as a phototrophic organism. This may be due to the fact that *Chloroflexus* contains chlorosomes (Fig. 2d), light-harvesting structures that are extremely effective in absorbing light at very low intensities (Oelze and Golecki, 1995). Indeed, the 2% of incident light intensity that supports photosynthetically active *Chloroflexus* in Octopus Spring may actually be greater than that needed to maintain natural populations of this organism.

The actual amount of light needed for growth of natural populations of *Chloroflexus* remains unknown. However, this organism could be an excellent experimental model for resolving the minimum light requirements for photosynthesis in anoxygenic phototrophs. Thus far, the lowest light intensities shown to support natural populations of anoxyphototrophs have been recorded for a brown-pigmented *Chlorobium* species that inhabits the depths of the Black Sea (Overmann et al., 1992). Like *Chloroflexus*, *Chlorobium* contains chlorosomes (see Fig. 2c). But the whole problem of light adaptation by phototrophic organisms naturally lends itself to an experimental setting like that of *Chloroflexus* in hot springs, and it is predicted that continued study of the effect of light intensity on pure cultures and natural populations of *C. aurantiacus* could reveal key physiological mechanisms for light adaptation that may have applications to the understanding of photosynthetic processes at high light intensities in green plants.

4.2. Autotrophy in Natural Populations of *Chromatium tepidum*

In the Mammoth area of Yellowstone, thin mats of the thermophilic purple sulfur bacterium *Chromatium* (*Thermochromatium*, Imhoff et al., 1998) *tepidum* develop in neutral pH thermal springs at temperatures between 40 and 60°C (see Castehnholz and Pierson, 1995, and Ward et al., 1989). Microbiological studies have shown that *Chloroflexus* and various chemoorganotrophic bacteria also inhabit these mats (Castenholz and Pierson, 1995; Ward et al., 1989, 1997). Because purple sulfur bacteria like *C. tepidum* can grow both photoautotrophically (light as the energy source and CO_2 as the carbon source) and photoheterotrophically (light as the energy source and organic carbon as the carbon source) (Madigan, 1986), the question emerged as to which nutritional pattern occurs in natural populations. In collaboration with geochemists, I approached this problem using carbon stable isotope technology to dissect the nutritional status of *C. tepidum* cells growing *in situ*.

4.2.1. Stable Carbon Isotope Methods

To follow the ensuing story, a short primer on stable carbon isotope technology and the concept of isotope fractionation is necessary. Two major stable (nonradioactive) isotopes of carbon exist, ^{12}C and ^{13}C; the natural abundances of ^{12}C and ^{13}C are roughly 95:5 (Coplen et al., 1983). For complex reasons that need not be described here, enzymes that "fix" CO_2, such as the Calvin cycle enzyme ribulose 1,5-bisphosophate carboxylase (RubisCO), discriminate against the heavier isotope of carbon ($^{13}CO_2$). Therefore, cell material produced by the action of RubisCO is enriched in ^{12}C and depleted in ^{13}C relative to the CO_2 source. This discrimination against the heavier isotope is referred to as isotope fractionation. Thus, carbon from laboratory cultures of *C. tepidum* grown autotrophically is isotopically "lighter" than carbon from cells grown on an organic carbon source like sodium acetate, where isotope fractionation does not occur (Madigan et al., 1989; see also Table 3). Note in Table 3 how enrichment of ^{12}C leads to depletion, relative to the standard, of ^{13}C, and thus the more *negative* the numbers for $\delta^{13}C$, the fractionation factor, the greater the depletion of ^{13}C.

Table 3
Carbon Isotope Fractionation by Pure Cultures and Natural Populations of the Thermophilic Purple Sulfur Bacterium *Chromatium tepidum*

I. Pure culture studies	Photoautotrophic[a]		Photoheterotrophic[a]
RubisCO[b]	33		0.1
Fractionation factors (‰)[c]			
Cells	−19.80		−7.61
Bacteriochlorophyll *a*	−19.27		−7.57
Phytol	−22.89		−12.42
Spirilloxanthin	−23.40		−12.87
Rhodovibrin	ND[e]		−13.13
II. Field studies	**Roland's Well**	**New Pit**	**New Spring**
Spring characteristics			
Temperature (°C)	52	55	53
pH	7.0	6.5	6.5
Inorganic carbon (mM)[d]	21.7	15.2	10.7
H_2S (μM)	41	34	ND[e]
Fractionation factors (‰)[c]			
Cells	−21.24	ND[e]	ND[e]
Bacteriochlorophyll *a*	−20.30	−19.59	−19.59
Phytol	−23.81	−22.85	−22.30
Spirilloxanthin	ND[e]	ND	−22.08
Rhodovibrin	ND[d]	−19.67	−21.01
Bacteriochlorophyll *c*	−21.7	−23.3	ND

[a]Laboratory cultures grown either photoautotrophically (CO_2/HCO_3^- as sole C source) or photoheterotropically on acetate/CO_2 (Madigan, 1986).
[b]Ribulose 1,5-bisphosphate carboxylase activity (nmol · min^{-1} · mg $protein^{-1}$) in crude cell extracts (Madigan et al., 1989).
[c]Isotopic composition ($\delta^{13}C$) $= \frac{{}^{13}C/{}^{12}C_{product} - {}^{13}C/{}^{13}C_{standard}}{{}^{13}C/{}^{12}C_{standard}}$

Fractionation factor (‰) $= [(\delta^{13}C_{product} + 1000)/(\delta^{13}C_{source} + 1000) - 1] \times 1000$ versus dissolved CO_2 (Coplen et al., 1983).
[d]Sum of H_2CO_3, HCO_3^-, and CO_3^{2-}.
[e]ND, not determined.

4.2.2. Carbon Isotope Fractionation by *Chromatium tepidum*

Stable isotope methods were applied to natural populations of *C. tepidum* to determine if cells, as hypothesized, grew photoautotrophically at the expense of geothermal H_2S and CO_2 emitted from their thermal spring habitats (Fig. 4). Before field work was begun, stable isotope studies on pure cultures of *C. tepidum* were necessary to validate the method. In such studies, fractionation factors for carbon were obtained exactly as would be expected of cells that fixed CO_2 by RubisCO; by contrast, organic carbon (in this case, sodium acetate) strongly repressed *C. tepidum* RubisCO, and thus cells were isotopically heavier (Table 3). Next, we prepared to take these methods to the field. However, because natural populations of *C. tepidum* were obviously not pure cultures, it was necessary that we be able to measure the $^{12}C/^{13}C$ content of cells and also that of specific biomarkers that could be reliably associated with this organism. The biomarkers chosen included the pigments bacteriochlorophyll *a* and rhodovibrin, among others (Table 3), and the isotopic composition of these materials purified from natural populations was determined along with that of intact cells of *C. tepidum* collected directly from mats.

Table 3 shows the results of these experiments and includes physical data on the three Mammoth thermal springs studied. As shown, the isotopic composition of natural populations of *C. tepidum* cells and of biomarkers extracted and purified from cells matched that of strictly photoautotrophically grown laboratory cultures. This fact strongly implies that *C. tepidum* is a true primary producer in its Yellowstone hot spring habitat. Moreover, it is likely that heterotrophic microorganisms that live in *C. tepidum* mats (of which many physiological groups are likely present, David M. Ward, personal communication, and Ward et al., 1997) and even the phototroph *Chloroflexus* (which grows very well as a photoheterotroph; see Madigan et al., 1974; Pierson and Castenholz, 1974a) obtain some or all of their carbon from organic compounds excreted by *Chromatium tepidum*. In fact, support for this hypothesis concerning *Chloroflexus* came from the finding that the isotopic composition of bacteriochlorophyll *c* extracted from natural populations of *Chloroflexus* was isotopically much lighter (Table 3) than carbon from autotrophically grown pure cultures (*Chloroflexus* uses the hydroxypropionate pathway instead of the Calvin cycle for CO_2 fixation and fractionates to a lesser degree during autotrophic growth than Calvin cycle organisms; Sirevåg et al., 1977). Thus, a relatively short food chain exists in *Chromatium tepidum* mats beginning with photoautotrophy by *C. tepidum* (driven by sunlight, geothermal H_2S and CO_2), followed by the likely excretion of organic compounds to its heterotrophic and photoheterotrophic (for example, *Chloroflexus*) neighbors. Further support for this hypothesis was obtained when analyses of spring water from the Mammoth springs studied showed that they were essentially devoid of organic matter (Madigan et al., 1989).

4.2.3. Stable Isotope Technology in Future Hot Spring Studies

Stable isotope methods could probably be adapted to study carbon flow in other Yellowstone thermal springs. Now that the microbial composition of alkaline hot spring mats is beginning to be understood (see other chapters in this volume and Ward et al., 1997), specific biomarkers could probably be identified that would allow for stable carbon isotope analyses of specific components of these structurally more complex microbial mats. One could also combine stable isotope technology with light adaptation studies like those described previously to follow CO_2 fixation by either *Chloroflexus* or *Synechococcus* during experimental perturbations of light intensity. Such studies have the potential to reveal the contributions of *Chloroflexus* or *Synechococcus* to net photosynthesis in the mat under different light regimes.

5. CONCLUSIONS

Undoubtedly many new thermophilic phototrophic microorganisms await discovery in Yellowstone. However, as this chapter has shown, thermophilic anoxygenic phototrophs are already a diverse group, and their photosynthetic activities may be of great importance to the Yellowstone ecosystem. Yellowstone's natural microbial laboratory offers an unparalleled opportunity to study phototrophic processes *in situ*. Especially interesting in this connection will be continued study of *Chloroflexus*; the unique photosynthetic properties (Strauss and Fuchs, 1993) and phylogenetic position (Gibson et al., 1985; Woese, 1987) of this organism strongly suggest that it has much more to tell us about the early evolution of photosynthesis than we know now. Continued study of *Chromatium tepidum* is also desirable, especially as regards the ecological role of *Chromatium*-like organisms (*Chromatium tepidum*?) observed in alkaline hot spring mats (Castenholz, 1977; Castenholz and Pierson, 1995). Finally, the anoxyphototroph *H. modesticaldum* (Kimble et al., 1995) requires further study to define its ecological role in photosynthesis in Yellowstone thermal areas. In this connection, *H. modesticaldum* is of special interest because it is a thermophilic nitrogen-fixing bacterium and is the only thermophilic anoxyphototroph known to produce endospores (Ormerod et al., 1990, 1996).

ACKNOWLEDGMENTS. Research on thermophilic anoxyphototrophs in the author's laboratory and in Yellowstone field studies has been funded by the National Science Foundation and the U.S. Department of Agriculture. The author thanks the staff of Yellowstone National Park, especially Mr. Robert Lindstrom, for permission to do research in Yellowstone and for many helpful suggestions.

REFERENCES

Amesz, J. 1995. The antenna-reaction center complex of heliobacteria. In Blankenship, R. E., Madigan, M. T., and Bauer, C. E. (eds.), *Anoxygenic photosynthetic bacteria* (pp. 687–697). Dordrecht, The Netherlands: Kluwer Academic.

Bauld, J., and Brock, T. D. 1973. Ecological studies of *Chloroflexis*, a gliding photosynthetic bacterium. *Arch. Mikrobiol.* **92:**267–284.

Blankenship, R. E., Madigan, M. T., and Bauer, C. E. (eds.). 1995. *Anoxygenic photosynthetic bacteria*. Dordrecht, The Netherlands: Kluwer Academic.

Brock, T. D. 1978. *Thermophilic microorganisms and life at high temperatures*. New York: Springer-Verlag.

Brock, T. D., and Brock, M. L. 1969. Effect of light intensity on photosynthesis by thermal algae adapted to natural and reduced sunlight. *Limnol. Oceanogr.* **14:**334–341.

Castenholz, R. W. 1969. Thermophilic blue-green algae and the thermal environment. *Bacteriol. Rev.* **33:**476–504.

Castenholz, R. W. 1977. The effect of sulfide on the blue-green algae of hot springs II. Yellowstone Park. *Microbial Ecol.* **3:**79–105.

Castenholz, R. W., Bauld, J., and Jørgensen, B. B. 1990. Anoxygenic microbial mats of hot springs: Thermophilic *Chlorobium* sp. *FEMS Microbiol. Ecol.* **74:**325–336.

Castenholz, R. W., and Pierson, B. K. 1995. Ecology of thermophilic anoxygenic phototrophs. In Blankenship, R. E., Madigan, M. T., and Bauer, C. E. (eds.), *Anoxygenic photosynthetic bacteria* (pp. 87–103). Dordrecht, The Netherlands: Kluwer Academic.

Charnock, C., Refseth, U. H., and Sirevåg, R. 1992. Malate dehydrogenase from *Chlorobium vibrioforme*, *Chlorobium tepidum*, and *Heliobacterium gestii*: Purification, characterization, and investigation of dinucleotide binding by dehydrogenases by use of empirical methods of protein sequence analysis. *J. Bacteriol.* **174:**1307–1313.

Coplen, T. B., Kendall, C., and Hopple, J. 1983. Comparison of stable isotope reference samples. *Nature* **302:** 236–238.

Doemel, W. N., and Brock, T. D. 1974. Bacterial stromatolites: Origin of laminations. *Science* **184:**1083–1085.

Doemel, W. N., and Brock, T. D. 1977. Structure, growth and decomposition of laminated algal-bacterial mats in alkaline hot springs. *Appl. Environ. Microbiol.* **34:**433–452.

Favinger, J., Statwald, R., and Gest, H. 1989. *Rhodospirillum centenum*, sp. nov., a thermotolerant cyst-forming anoxygenic photosynthetic bacterium. *Antonie Leeuwenhoek* **31:**317–322.

Ferris, M. J., and Ward, D. M. 1997. Seasonal distribution of dominant 16S rRNA-defined populations in a hot spring microbial mat examined by denaturing gel electrophoresis. *Appl. Environ. Microbiol.* **63:**1375–1381.

Frigaard, N-U., Taikaichi, S., Hirota, M., Shimada, K., and Matsuura, K. 1997. Quinones in chlorosomes of green sulfur bacteria and their role in the redox-dependent fluorescence studied in chlorosome-like bacteriochlorophyll *c* aggregates. *Arch. Microbiol.* **167:**343–349.

Garcia, D., Parot, P., Verméglio, A., and Madigan, M. T. 1986. The light-harvesting complexes of a thermophilic purple sulfur photosynthetic bacterium *Chromatium tepidum*. *Biochim. Biophys. Acta* **850:**390–395.

Gibson, J., Ludwig, W., Stackebrant, E., and Woese, C. R. 1985. The phylogeny of the green photosynthetic bacteria: Absence of a close relationship between *Chlorobium* and *Chloroflexus*. *Syst. Appl. Microbiol.* **6:**152–156.

Giovannoni, S. J., Revsbech, N. P., Ward, D. M., and Castenholz, R. W. 1987. Obligately phototrophic *Chloroflexus*: Primary production in anaerobic hot spring microbial mats. *Arch. Microbiol.* **147:**80–87.

Halfen, L. N., Pierson, B. K., and Francis, C. W. 1972. Carotenoids of a gliding organism containing bacteriochlorophylls. *Arch. Microbiol.* **82:**240–246.

Heda, G. D., and Madigan, M. T. 1985. Thermal properties and oxygenase activity of ribulose-1,5-bisphosphate carboxylase from the thermophilic purple bacterium *Chromatium tepidum*. *FEMS Microbiol. Lett.* **51:**45–50.

Heda, G. D., and Madigan, M. T. 1986a. Utilization of amino acids and lack of diazotrophy in the thermophilic anoxygenic phototroph *Chloroflexus aurantiacus*. *J. Gen. Microbiol.* **132:**2469–2473.

Heda, G. D., and Madigan, M. T. 1986b. Aspects of nitrogen fixation in relationship *Chlorobium*. *Arch. Microbiol.* **143:**330–336.

Heda, G. D., and Madigan, M. T. 1989. Purification and characterization of the thermostable ribulose-1,5-bisphosphate carboxylase/oxygenase from the thermophilic purple bacterium *Chromatium tepidum*. *Eur. J. Biochem.* **184:**313–319.

Holo, H., and Sirevåg, R. 1986. Autotrophic growth and CO_2 fixation of *Chloroflexus aurantiacus*. *Arch. Microbiol.* **145:**173–180.

Imhoff, J. F., Süling, J., and Petri, R. 1998. Phylogenetic relationships among the Chromatiaceae, then taxonomic reclassification and description of the new genera *Allochromatium*, *Halochromatium*, *Isochromatium*, *Marichromatium*, *Thiococcus*, *Thiohalocapsa*, and *Thermochromatium*. *Int. J. Syst. Bacteriol.* **48:**1129–1143.

Kimble, L. K., Mandelco, L., Woese, C. R., and Madigan, M. T. 1995. *Heliobacterium modesticaldum*, sp. nov., a thermophilic heliobacterium of hot springs and volcanic soils. *Arch. Microbiol.* **163:**259–267.

Kimble, L. K., Stevenson, A. K., and Madigan, M. T. 1994. Chemotrophic growth of heliobacteria in darkness. *FEMS Microbiol. Lett.* **115:**51–56.

Kondratieva, E. N., Pfennig, N., and Trüper, H. G. 1992. The phototrophic prokaryotes. In Balows, A., Trüper, H. G., Dworkin, M., Harder, W., and Schleifer, K-H. (eds.), *The Prokaryotes* (2nd ed., pp. 312–330). New York: Springer-Verlag.

Madigan, M. T. 1984. A novel photosynthetic purple bacterium isolated from a Yellowstone hot spring. *Science* **225:**313–315.

Madigan, M. T. 1986. *Chromatium tepidum* sp. n., a thermophilic photosynthetic bacterium of the family *Chromatiaceae*. *Int. J. Syst. Bacteriol.* **36:**222–227.

Madigan, M. T. 1995. Microbiology of nitrogen fixation by anoxygenic photosynthetic bacteria. In Blankenship, R. E., Madigan, M. T., and Bauer, C. E. (eds.), *Anoxygenic photosynthetic bacteria* (pp. 915–928). Dordrecht, The Netherlands: Kluwer Academic.

Madigan, M. T., and Brock, T. D. 1975. Photosynthetic sulfide oxidation by *Chloroflexus aurantiacus*, a filamentous, photosynthetic gliding bacterium. *J. Bacteriol.* **122:**782–784.
Madigan, M. T., and Brock, T. D. 1977. Adaptation by hot spring phototrophs to reduced light intensities. *Arch. Microbiol.* **113:**111–120.
Madigan, M. T., Peterson, S. R., and Brock, T. D. 1974. Nutritional studies on *Chloroflexus*, a filamentous, photosynthetic, gliding bacterium. *Arch. Microbiol.* **100:**97–103.
Madigan, M. T., Takigiku, R., Lee, R. G., Gest, H., and Hayes, J. M. 1989. Carbon isotope fractionation by thermophilic phototrophic sulfur bacteria: Evidence for autotrophic growth in natural populations. *Appl. Environ. Microbiol.* **55:**639–644.
Miller, K. R., Jacob, J. S., Smith, U., Kolaczkowski, S., and Bowman, M. K. 1986. *Heliobacterium chlorum*: Cell organization and structure. *Arch. Microbiol.* **146:**111–114.
Nozawa, T., Fukada, T., Hatano, M., and Madigan, M. T. 1986. Organization of intracytoplasmic membranes in a novel thermophilic purple photosynthetic bacterium as revealed from absorption, circular dichroism, and emission spectra. *Biochim. Biophys. Acta* **852:**191–197.
Nozawa, T., and Madigan, M. T. 1991. Temperature and solvent effects on reaction centers from *Chloroflexus aurantiacus* and *Chromatium tepidum*. *J. Biochem.* **110:**588–594.
Oelze, J., and Golecki, J. R. 1995. Membranes and chlorosomes of green bacteria: Structure, composition and development. In Blankenship, R. E., Madigan, M. T., and Bauer, C. E. (eds.), *Anoxygenic photosynthetic bacteria* (pp. 259–278). Dordrecht, The Netherlands: Kluwer Academic.
Ormerod, J. G., Kimble, L. K., Nesbakken, T., Torgersen, Y. A., Woese, C. R., and Madigan, M. T. 1996. *Heliophilum fasciatum* gen. nov. and sp. nov., and *Heliobacterium gestii* sp. nov.: Endospore-forming heliobacteria from rice field soils. *Arch. Microbiol.* **166:**226–234.
Ormerod, J. G., Nesbakken, T., and Torgersen, T. 1990. Phototrophic bacteria that form heat resistant endospores. In Baltscheffsky, M. (ed.), *Current research in photosynthesis* (Vol. 4, pp. 935–938). Dordrecht, The Netherlands: Kluwer Academic.
Overmann, J., Cypionka, H., and Pfennig, N. 1992. An extremely low-light-adapted phototrophic sulfur bacterium from the Black Sea. *Limnol. Oceanogr.* **370:**150–155.
Pierson, B. K., and Castenholz, R. W. 1971. Bacteriochlorophylls in gliding filamentous prokaryotes of hot springs. *Nature New Biol.* **233:**25–27.
Pierson, B. K., and Castenholz, R. W. 1974a. A phototrophic gliding filamentous bacterium of hot springs, *Chloroflexus aurantiacus*, gen. and sp. nov. *Arch. Microbiol.* **100:**5–24.
Pierson, B. K., and Castenholz, R. W. 1974b. Studies of pigments and growth in *Chloroflexus aurantiacus*, a phototrophic filamentous bacterium. *Arch. Microbiol.* **100:**283–305.
Pierson, B. K., and Castenholz, R. W. 1995. Taxonomy and physiology of filamentous anoxygenic phototrophs. In Blankenship, R. E., Madigan, M. T., and Bauer, C. E. (eds.), *Anoxygenic photosynthetic bacteria* (pp. 31–47). Dordrecht, The Netherlands: Kluwer Academic.
Pierson, B. K., Giovannoni, S. J., Stahl, D. A., and Castenholz, R. W. 1985. *Heliothrix oregonensis*, gen. nov., sp. nov., a phototrophic filamentous gliding bacterium containing bacteriochlorophyll *a*. *Arch. Microbiol.* **142:** 164–167.
Postgate, J. R. 1982. *The fundamentals of nitrogen fixation*. Cambridge, England: Cambridge University Press.
Redburn, A. C., and Patel, B. K. C. 1993. Phylogenetic analysis of *Desulfotomaculum thermobenzoicum* using polymerase chain reaction-amplified 16S rRNA-specific DNA. *FEMS Microbiol. Lett.* **113:**81–86.
Resnick, S. M., and Madigan, M. T. 1989. Isolation and characterization of a mildly thermophilic nonsulfur purple bacterium containing bacteriochlorophyll *b*. *FEMS Microbiol. Lett.* **65:**165–170.
Ruff-Roberts, A. L., Kuenen, G. J., and Ward, D. M. 1994. Distribution of cultivated and uncultivated cyanobacteria and *Chloroflexus*-like bacteria in hot spring microbial mats. *Appl. Environ. Microbiol.* **60:**697–704.
Sirevåg, R., Buchanan, B. B., Berry, J. A., and Troughton, J. H. 1977. Mechanisms of CO_2 fixation in bacterial photosynthesis studied by the carbon isotope fractionation technique. *Arch. Microbiol.* **112:**35–38.
Sirevåg, R., and Castenholz, R. W. 1979. Aspects of carbon metabolism in *Chloroflexus*. *Arch. Microbiol.* **120:** 151–153.
Stadtwald-Demchick, R., Turner, F. R., and Gest, H. 1990. *Rhodopseudomonas cryptolactis*, sp. nov., a new purple bacterium. *FEMS Microbiol. Lett.* **71:**117–122.
Stevenson, A. K., Kimble, L. K., Woese, C. R., and Madigan, M. T. 1997. Characterization of new phototrophic heliobacteria and their habitats. *Photosynthesis Res.* **53:**1–12.
Stewart, W. D. P. 1970. Nitrogen fixation by blue-green algae in Yellowstone thermal areas. *Phycologia* **9:**261–268.

Strauss, G., and Fuchs, G. 1993. Enzymes of a novel autotrophic CO_2 fixation pathway in the phototrophic bacterium *Chloroflexus aurantiacus*. *Eur. J. Biochem.* **215**:633–643.

Synstad, B., Emmerhoff, O., and Sirevåg, R. 1996. Malate dehydrogenase from the green gliding bacterium *Chloroflexus aurantiacus* is phylogenetically related to lactic dehydrogenases. *Arch. Microbiol.* **165**:346–353.

Taikaichi, S., Inoue, K., Akaike, M., Kobayashi, M., Oh-oka, H., and Madigan, M. T. 1997a. The major carotenoid in all known species of heliobacteria is the C_{30} carotenoid 4,4′-diaponeurosporene, not neurosporene. *Arch. Microbiol.* **168**:277–281.

Taikaichi, S., Tsuji, K., Matsuura, K., and Shimada, K. 1995. A monocyclic carotenoid glucoside ester is a major carotenoid in the green filamentous bacterium *Chloroflexus aurantiacus*. *Plant Cell Physiol.* **36**:773–778.

Taikaichi, S., Wang, Z-Y., Unetsu, M., Nozawa, T., Shimada, K., and Madigan, M. T. 1997b. New carotenoids from the thermophilic green sulfur bacterium *Chlorobium tepidum*: 1′,2′-Dihydro-β-carotene, 1′-2′-dihydrochlorobactene and OH-chlorobactene glucoside ester, and the carotenoid composition of different strains. *Arch. Microbiol.* **168**:270–276.

Wahlund, T. M., Woese, C. R., Castenholz, R. W., and Madigan, M. T. 1991. A thermophilic green sulfur bacterium from New Zealand hot spring, *Chlorobium tepidum* sp. nov. *Arch. Microbiol.* **156**:81–90.

Wahlund, T. M., and Madigan, M. T. 1993. Nitrogen fixation by the thermophilic green sulfur bacterium, *Chlorobium tepidum*. *J. Bacteriol.* **175**:474–478.

Ward, D. M., Bateson, M. M., Weller, R., and Ruff-Roberts, A. L. 1992. Ribosomal RNA analysis of microorganisms as they occur in nature. *Adv. Microb. Ecol.* **12**:219–286.

Ward, D. M., Weller, R., Shiea, J., Castenholz, R. W., and Cohen, Y. 1989. Hot spring microbial mats—anoxygenic and oxygenic mats of possible evolutionary significance. In Cohen, Y., and Rosenberg, E. (eds.), *Microbial mats—physiological ecology of benthic microbial communities* (pp. 3–15). Washington, D.C.: American Society for Microbiology.

Ward, D. M., Santegoeds, C. M., Nold, S. C., Ramsing, N. B., Ferris, M. J., and Bateson, M. M. 1997. Biodiversity within hot spring microbial mat communities: Molecular monitoring of enrichment cultures. *Antonie Leeuwenhoek* **71**:143–150.

Woese, C. R. 1987. Bacterial evolution. *Microbiol. Rev.* **51**:221–271.

10

Algal Physiology at High Temperature, Low pH, and Variable pCO_2

Implications for Evolution and Ecology

Lynn J. Rothschild

1. INTRODUCTION: WHY THE MICROBIAL MATS OF YELLOWSTONE

Modeling ancient ecosystems challenges the experimental biologist. Through both visual and chemical analysis, the fossil record provides invaluable information on past community composition and hints at the physiology of the members of the ecosystem. The molecular fossil record, as revealed through DNA sequence analysis, helps identify modern relatives of ancient communities. But to more fully understand ancient ecosystems, experimental work on modern analogs (actualistic paleontology) is essential. Furthermore, modern analogs of ancient ecosystems must be studied under appropriate physical conditions. The earth's environment has changed considerably in the 3.5 or so billion years since life arose. Atmospheric levels of inorganic carbon (C_i) were probably higher, and free oxygen was rare. Surficial fluxes of solar UVB (280 to 320 nm) radiation were probably substantially higher, and day length was shorter. Ocean temperature and pH may have been similar to modern conditions. Interestingly, some of the same environmental factors are again changing, albeit modestly compared to the shifts that have occurred since the Precambrian. Atmospheric pCO_2 is rising primarily as a by-product of fossil fuel combustion, and surficial UVB fluxes are rising as a result of the depletion of stratospheric ozone. The associated phenomena of increasing temperature and acidity due to such factors as atmo-

Lynn J. Rothschild • Ecosystem Science and Technology Branch, NASA/Ames Research Center, Moffett Field, California 94035-1000.

Thermophiles: Biodiversity, Ecology, and Evolution, edited by Reysenbach *et al.* Kluwer Academic / Plenum Publishers, New York, 2001.

spheric deposition of industrial SO_2 emissions and drought are becoming problems for some temperate aquatic ecosystems (e.g., Yan et al., 1996; Gorham, 1996).

The acid springs and acid hot springs of Yellowstone National Park provide model ecosystems for studying the effects of low pH and high temperature on ecosystems. The acid springs are populated by microbes, the dominant organisms through much of the Precambrian. Eukaryotic algae are by far the most conspicuous inhabitants of many of these springs, and photosynthetic bacteria are absent because they do not live much below pH 5 (Doemel and Brock, 1971). Photosynthetic eukaryotes, it is thought, arose by at least 1.5 Ga, and possibly 2.1 Ga (e.g., Han and Runnegar, 1992), and they participated in stromatolite formation after that time (Walter et al., 1992). Outside of acid springs, eukaryotic algal-dominated mats are rare now and probably were rare during the Precambrian (Ward et al., 1989). Mats in which the biomass is dominated by eukaryotes provide models for a eukaryotic response to selected environmental conditions because they represent virtually a monoculture that has only a minor component of prokaryotes or more recently evolved eukaryotes. In such systems, any primary productivity or DNA synthesis can be reasonably attributed to the algae.

Microbial mats are found in Norris Geyser Basin in the western part of Yellowstone National Park (Fig. 1). Norris is the hottest, most volatile thermal area in North America, perhaps on earth. Many of the hot springs in Norris, including those studied here, are acidic. Because of the low pH, the concentration of dissolved C_i (DIC) is also extremely low. Thus, microbial mats there provide ecosystems in which to study the effects of carbon limitation.

The taxonomic composition of the mats is useful for studies of the Precambrian. One of the mats studied here was composed primarily of the unicellular red alga, *Cyanidium caldarium* and the other of the green alga *Zygogonium*. Green and red algae are probably among the most ancient lineages of photosynthetic eukaryotes (e.g., Rothschild and Heywood, 1988; Knoll, 1992; Ragan and Gutell, 1995). Fossil evidence suggests that multicellular red algae emerged at least 1.25 Ga (Butterfield et al., 1990), and unicellular genera undoubtedly predated the multicellular forms. *Cyanidium* is considered primitive for ultrastructural and biochemical reasons (Seckbach and Ott, 1994). The physiology of

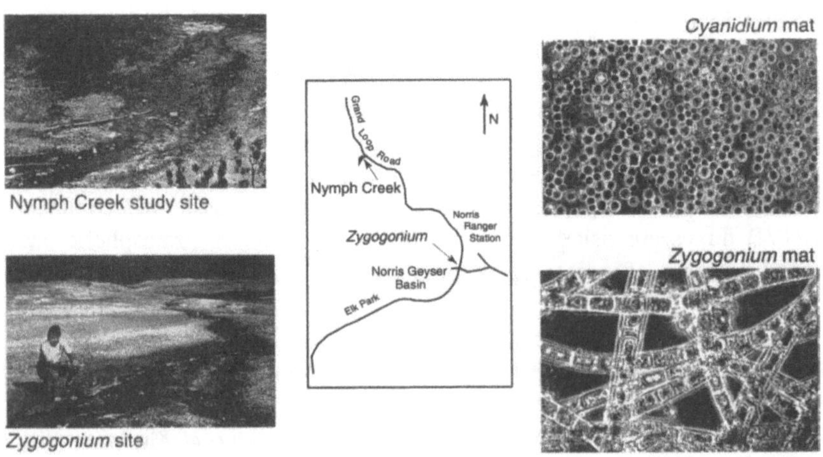

Figure 1. Study sites. For a more complete description of the field sites, see Brock (1978).

Cyanidium, including its ability to grow in relatively high temperature, low pH, and pure CO_2, suggest that it may have evolved early (Ott and Seckbach, 1994). Besides its occurrence in Yellowstone, *Cyanidium* is widely distributed in acidic thermal areas (e.g., Brock, 1978; Gromov et al., 1979; Gerasimenko and Miller, 1985).

This chapter offers an overview of my research on the physiology of two algal communities that live in acid springs and acid hot springs in Yellowstone. The aims of this work were to understand the functioning of these communities and to use them as analog ecosystems for assessing the effects of low pH, high temperature, and low and high concentrations of DIC on proposed ancient and modern ecosystems that face the influences of global change. Physiological features emphasized were primary productivity and DNA synthesis for the following reasons: (1) Ecosystem dynamics begins with the production of organic carbon from inorganic carbon. (2) The rates of this primary production affect the rate at which organic carbon is excreted by primary producers, organismal growth rates, and other ecosystem processes. Measuring subsequent fluxes of carbon can elucidate the fate of this fundamental building element of organic compounds. DNA synthesis rates can indicate growth and environmental damage that can lead to mutation.

2. MATERIALS AND METHODS

2.1. Description of Organisms

The *Cyanidium caldarium* studies reported here were conducted in Nymph Creek, an acidic (pH ~2.8 to 3.3) hot spring that flows into Nymph Lake (Fig. 1). The primary bacterial component in Nymph Creek was an acidophilic strain of *Bacillus coagulans*, and the primary fungal component was *Dactylaria gallopava* (Belly et al., 1973). The water temperature of the study site remained almost constant during the day; it varied between 39.3 and 41.6°C on 22 September 1995, and at the same time the air temperature varied between −2.2 and 22°C. Similarly, water temperature at the study site in Nymph Creek varied only between 40 and 44°C on 24 June 1993, a day that began with light snow (Fig. 2).

There were dark purple mats of a species of *Zygogonium*, a filamentous green alga of the family Zygnemataceae, in shallow acidic areas adjacent to thermal features (Fig. 1). The purple color comes from vacuolar pigments, likely iron-tannin complexes (Alston, 1958). Microscopic examination revealed some unicellular algae among the filaments of *Zygogonium*, including moderate to large numbers of *Euglena mutabilis* and sometimes small numbers of *Chlorella* sp. and *Chlamydomonas acidophila*. The *Zygogonium* studied was found at ambient temperatures in an acidic (pH ~2.2–3.3) stream near Norris Annex, south of Nymph Creek and just north of Norris Junction (Fig. 1). The temperature of the mat varied with the air temperature during the day. For example, on 22 September 1995, the temperature on the surface of the mat varied from 4.2°C at 7:45 to a high of 19.7°C at 14:30. Diurnal temperate data for June are shown in Fig. 2.

2.2. Primary Productivity

The ^{14}C incorporation method was used to determine primary productivity. In this method, production of acid stable material (organic carbon) from inorganic carbon is measured (Strickland and Parsons, 1968; Peterson, 1980). Replicate intact cores (17 cm^2) of

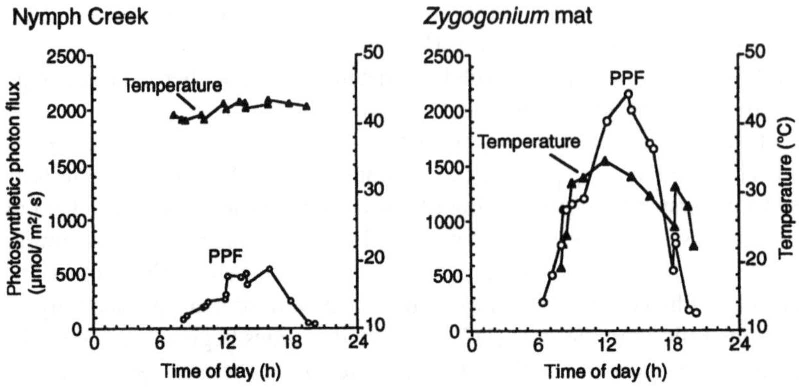

Figure 2. Diurnal variation in light and temperature at the study sites. Nymph Creek is a large, free-flowing acid hot spring leading to Nymph Lake. It is fed by a series of small hot springs and maintains a fairly constant temperature. The area around the creek is wooded, and thus most of the creek does not receive full solar radiation. In contrast, the moist acidic areas of Norris Annex where the *Zygogonium* mat were found are open to full solar radiation that controls the temperature of the mat. Light (photosynthetic photon flux) is in μmoles photons/m²/s of photosynthetically active radiation (400 to 700 nm), and was determined by using a LiCor Model 189 quantum/radiometer/photometer.

the mats were placed in Whirl-pak® bags, plastic bags that are transparent to visible and UV radiation (Fig. 3). Spring water supplemented with 1 μCi/ mL $NaH^{14}CO_3$ (New England Nuclear NEC 086H) was added to the bags. The bags were sealed and returned to the stream. The samples were incubated for 15 or 20 min, placed in the dark on dry ice, and returned to the lab for analysis. In the lab, the samples were defrosted, 50 mL of deionized water was added, and the samples were homogenized by sonication. Duplicate aliquots (100 μL) were placed in scintillation vials, acidified with acetic acid, and dried under a stream of warm air. Samples were resuspended in 200 μL warm water, scintillation fluid (Ecolume, ICN Biomedicals, Inc.) was added, and counts were accumulated by scintillation spectroscopy. To convert cpm to μg C fixed/m²/min, the following formula, based on Strickland and Parsons (1968) was used:

$$\mu\text{g C fixed/m}^2\text{/min} = \frac{(\text{dpm}_{\text{sample}}/\text{m}^2 - \text{dpm}_{\text{control}}/\text{m}^2)(\mu\text{g C/dpm})(1.05)}{\text{incubation time in minutes}} \quad (1)$$

where the dpm/m² was calculated from cpm/m² allowing for quench by using internal standards, the control was killed or did not have ^{14}C added, and μg C/dpm was calculated by knowing the concentration of C_i and the concentration of ^{14}C per mL incubation mix and using an atomic weight of 12.0 for carbon. The concentration of C_i was determined by acidifying the sample to pH <5 to convert all C_i to CO_2, then measuring total dissolved CO_2 using a commercially available CO_2 electrode (Ingold model # 15-232-3000). The standard factor of 1.05 was used to compensate for the discrimination of the carbon fixation enzyme, ribulose-1,5-bisphosphate carboxylase, against ^{14}C relative to ^{12}C.

To determine the effect of elevated levels of pCO_2 on photosynthesis, mats were incubated and processed as described before, but stream water was supplemented with $NaHCO_3$. To adjust the pH of the incubation mix, 0.1 N HCl was added until the desired pH was reached. To determine the effect of longer exposure of mats to elevated pCO_2, mats were placed in Whirl-pak® bags with either 5 mL stream water or water supplemented with 5 mM $NaHCO_3$ whose pH was lowered to approximately ambient levels (pH 3), and were replaced in the stream to incubate overnight. To conduct the experiment, the preincubation water was drained from the bag, and a freshly made incubation mix of stream water containing 1–2 μCi/mL $NaH^{14}CO_3$ with or without additional bicarbonate and additional nutrients (25 mM sodium acetate or 10 mM ammonia chloride) was added. The rest of the incubation was conducted, and the samples were processed as described before.

2.3. DNA Synthesis

Mat samples were placed in Whirl-pak® bags and an incubation mix of stream water containing 15 μCi/mL^{33}P-phosphate (New England Nuclear NEZ080) for a total of 5 mL. The samples were returned to the stream for 2 hours. Incubation was stopped by placing bags on dry ice in the dark. Sample bags were returned to the laboratory in that condition. DNA was extracted by using a modification of Doyle and Doyle (1990). Cells were ground to a powder in liquid N_2. The powder was transferred to a solution of 250 μL each of sample buffer (100 mM Tris, 5 mM EDTA) and CTAB isolation buffer (4% [w/v] (CTAB), 2.8 m NaCl, 0.4% [v/v] 2-mercaptoethanol, 40 mM EDTA, 200 mM Tris-HCl, [pH 8.0]). Samples were incubated at 60°C for 30 min with occasional gentle swirling and then were extracted once with an equal volume of chloroform:isoamyl alcohol (24:1 v:v). The aqueous phase was removed to a new tube, and nucleic acids were precipitated with 2/3 volume cold isopropanol and washed with 76% ethanol containing 10 mM ammonium acetate. The pellet was dried, resuspended in TE (10 mM Tris, 1 mM EDTA) and checked for purity spectrophotometrically (ratio of OD_{260} to $OD_{280} = 1.5 \pm 0.2$ and visually on a 0.8% agarose

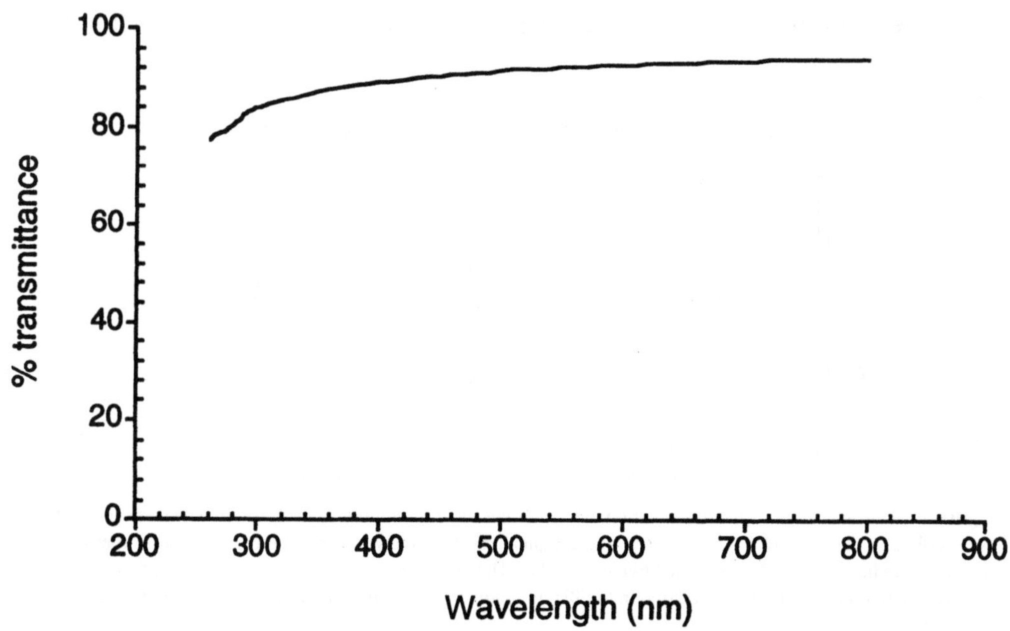

Figure 3. Spectral characteristics of Whirl-pak® bags.

gel. Little, if any, RNA was detected. DNA concentration was calculated from spectrophotometric data using the equations of Warburg and Christian (1942). Radioactive counts were determined from a 100 μL aliquot of the DNA.

2.4. Partitioning of Photosynthate into DNA

Mats were incubated in the presence of 20 μCi/mL $NaH^{14}CO_3$, in the same way as the carbon fixation experiments. The incubation period was 2 h, as in the DNA synthesis experiments, which were conducted concurrently. The DNA was purified in the lab as in the DNA synthesis experiments.

3. RESULTS

3.1. Primary Productivity

3.1.1. Diurnal Patterns of Carbon Fixation

The initial experiments were designed to elucidate natural diurnal patterns of carbon fixation in these mats. As shown in Fig. 4, carbon fixation increases in the morning, decreases at midday, shows a second peak in the afternoon, and decreases in the evening. This is similar to the pattern obtained in hypersaline and marine microbial mats (Bebout et al., 1987; Villbrandt et al., 1990; Rothschild, 1991), plankton-dominated lake communities (Peterson et al., 1977; Vincent, 1980), and a subtidal population of the red alga *Gracilaria verrucosa* (Hoffman and Dawes, 1980). The cause of this midday dip is not completely known. Possibly influential factors include temperature, pH, salinity, water stress, and solar radiation. The Yellowstone mats are unusual in that many of these factors are naturally constrained through the day, thus providing a controlled environment in which to test

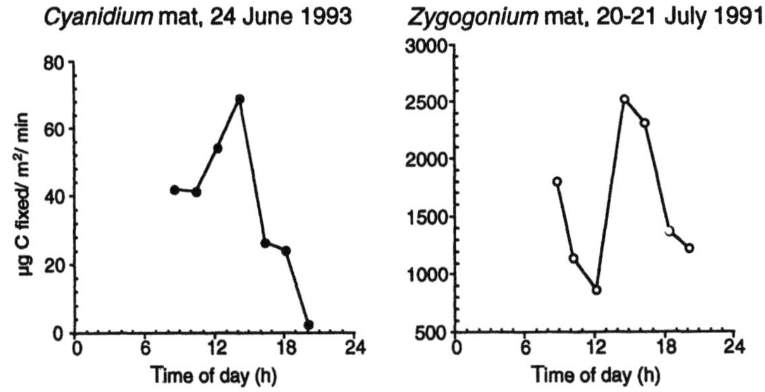

Figure 4. Diurnal patterns of carbon fixation in *Cyanidium* and *Zygogonium* mats. Experiments were conducted, as described in the text, over several years, and a typical result is shown in this figure. Triplicate experimental samples were incubated and subsampled twice in the lab. The values plotted are an average of these data. Note the pronounced midday dip in carbon fixation in the *Zygogonium* mat, similar to midday dips in other photosynthetic organisms.

hypotheses for the cause of the midday dip. Although temperature of the *Zygogonium* mats varies diurnally, the temperature in Nymph Creek varied less than a few degrees throughout the day (Fig. 2) because Nymph Creek is continuously fed by a series of small geysers. Thus, temperature is an unlikely cause for the midday dip. Diurnal variations in pH or salinity were not detected. Water stress, a factor that has been suggested as a cause of a midday dip in photosynthesis in angiosperms, cannot occur because the mats are submerged. Solar radiation was measured and peaks near the midday dip (Fig. 2). High levels of visible radiation are known as a cause of photoinhibition (reviewed in Powles, 1984) and may be responsible. The nutritional status of the cells, specifically a deficiency in fixed nitrogen by late morning, could also cause the dip. This argument suggests that a lack of fixed nitrogen limits the ability of the cells to convert fixed carbon to nitrogenous compounds such as proteins, resulting in down-regulation of carbon fixation until more fixed nitrogen becomes available.

3.1.2. Carbon Fixation under Elevated CO_2

Having established the natural diurnal pattern of carbon fixation, the next question was how elevated levels of C_i would affect this pattern. This is of interest to test the effect that low levels of DIC may have on the photosynthetic rate and because levels of DIC may have been substantially higher during the Archaean and, in some systems, may again be increasing (see Section 4.2, Ecological and Evolutionary Implications, for a more complete discussion of these issues). Diurnal experiments were repeated adding an additional 10 mM sodium bicarbonate to the incubation mix before adding the mix to the mat sample. Additional bicarbonate had an immediate and profound effect on carbon fixation rates in these mats (Fig. 5). This result is similar to previous results for marine and hypersaline cyanobacterial mats (Rothschild and Mancinelli, 1990; Rothschild, 1991, 1994b) but is even

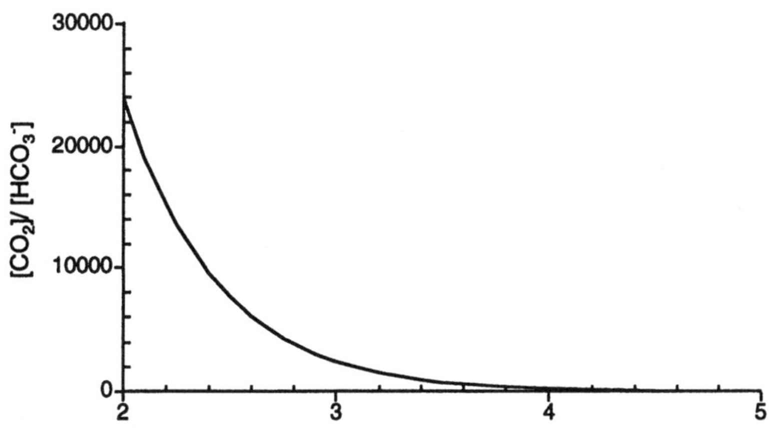

Figure 5. The effect of pH on carbonate equilibrium. At the pH of both the Nymph Creek and the *Zygogonium* mats, free CO_2 far exceeds the bicarbonate ion. The dissociation constants were taken from Buch (1960) and assume 0% salinity and a temperature of 25°C.

more dramatic. The effect of additional bicarbonate on the acidic Yellowstone mats is more profound because the ambient concentration of DIC is very low. For example, in 1992, DIC was about 0.23 mM for the water in Norris Annex where the *Zygogonium* lived and 0.09 mM DIC for Nymph Creek. In contrast, ocean water contains about 2 mM DIC. Thus, proportionally, an additional 10 mM bicarbonate is far more significant for the acidic than for the marine mats.

The stream water at both sites was very weakly buffered, so that when an additional 10 mM sodium bicarbonate was added to the water, the pH rose from less than 2.5 to more than 6. The higher pH had little or no effect on carbon fixation rates in *Cyanidium* mat from about noon to 1600 h but inhibited carbon fixation at other times of day, particularly when light levels were lower (Fig. 5). This suggests that *Cyanidium* uses a carbon acquisition mechanism at high pH, probably a bicarbonate pump, which uses energy that is otherwise diverted from carbon fixation or, alternatively, it could suggest that carbon fixation is carbon saturated with the additional 10 mM bicarbonate during the middle of the day, and thus other factors control carbon fixation rates. At the end of the day, carbon fixation is carbon limited at the high pH because a bicarbonate pump does not exist. It would seem odd if the latter were true because light should be limiting at the end of the day, so carbon should be relatively more abundant than from noon to 1600 h.

The data clearly showed an immediate and dramatic response to exposure to elevated levels of C_i, but would such a response continue during a more prolonged exposure? The literature on angiosperms suggests that when other environmental variables (e.g., water, nutrients, light) are limiting, the photosynthetic apparatus may acclimate to elevated levels of CO_2, thus counteracting the short-term fertilization effect of CO_2 (review in Long and Drake, 1992). For example, *in situ* studies of Arctic tundra exposed to doubled CO_2 showed a sixfold fertilization effect during the first growing season but acclimated within three years under ambient temperature (Oechel et al., 1994). Interestingly, the CO_2 fertilization effect persisted with increased temperature, suggesting that temperature limits the photosynthetic rate.

Long-term studies of wild algal communities, similar to those that have angiosperm communities, have not been attempted. But because reproductive rates of algae are substantially faster than those of plants, algae should acclimate at a much faster rate. To begin to study the question of longer term effects, *Cyanidium* and *Zygogonium* samples were exposed in June 1993 to enhanced levels of C_i for 24 hours and then tested for photosynthetic response. Cores (17 cm^2) were placed in sterile Whirl-pak® bags and 5 mL of either stream water or stream water supplemented with 5 mM $NaHCO_3^-$ whose pH was adjusted to 3.3 was added. The samples were replaced in the sampling location for 24 hours. Experiments were then conducted from 1700 to 2000 to measure carbon fixation rates at the preincubation levels of C_i with and without an additional source of fixed carbon (25 mM sodium acetate) or fixed nitrogen (10 mM ammonium chloride). Carbon fixation of both mats was enhanced more than an order of magnitude even after preincubation (Fig. 6). As in previous experiments (e.g., Rothschild, 1991; unpublished data from Yellowstone), the presence of an exogenous source of fixed carbon decreased carbon fixation rates, although the overall rate of carbon fixation was still enhanced in the elevated C_i treatments (Fig. 6). The presence of fixed nitrogen enhanced carbon fixation in both treatments, but again the relative fixation rate was higher in the samples preincubated in the presence of elevated C_i (Fig. 6).

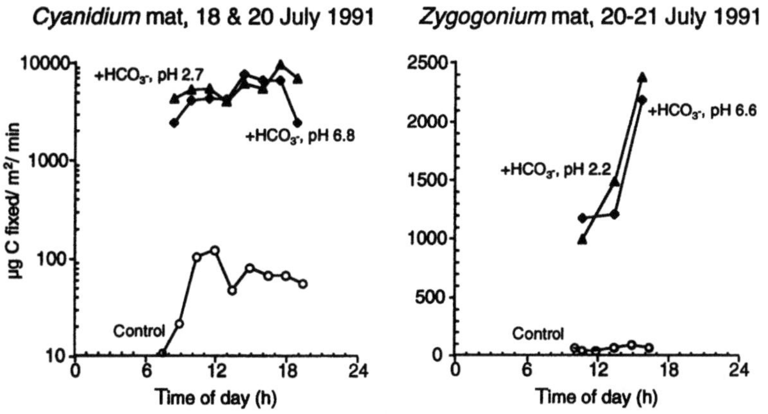

Figure 6. Effect of additional C_i on carbon fixation rates. When the samples are supplemented with an additional 10 mM bicarbonate, carbon fixation rates increase dramatically at both acidic and neutral pH. When 10 mM bicarbonate was added, the pH rose to >6. Experiments were conducted with samples at the high pH and on samples where the pH had been reduced to ambient levels by adding HCl. During the middle of the day, this difference in pH had no significant effect on carbon fixation rates.

3.2. DNA Synthesis

A study of *in situ* rates of DNA synthesis was initiated because so little is known about this fundamental physiological function in nature and because it is indicative of cell activity, especially growth rate. Instantaneous growth rates would be impossible to obtain on a filamentous mat by cell counts. DNA synthesis rates are routinely determined by studying the incorporation of a radiolabeled nucleotide (e.g., [^{3}H]thymidine) into nucleic acids (Fuhrman and Azam, 1980; Thorn and Ventullo, 1988). This method is often used to estimate cell growth rates based on the assumption that DNA synthesis rates are correlated with cell division rates. However, this method has several problems associated with it (review in Staley and Konopka, 1985). Eukaryotes, in general, do not incorporate thymidine (Staley and Konopka, 1985). In some natural microbial assemblages, only the heterotrophic members take it up (Carman, 1990), and sulfate-reducing bacteria may not incorporate it either (Gilmour et al., 1990). Some organisms metabolize the thymidine rather than incorporate it into DNA (e.g., Jeffrey et al., 1990; Wellsbury et al., 1993; Rothschild, unpublished observations on hypersaline cyanobacterial mats). The resulting "non-specific" labeling of RNA and protein may be as high as 98% of the incorporated label (Brittain and Karl, 1990; Jeffrey et al., 1990).

During the last few years, I developed a method to measure the *in situ* incorporation rates of [^{33}P]phosphate into DNA in a manner similar to the approach used by Winn and Karl (1986) on marine plankton. Using this method, apparent rates of DNA synthesis in both *Zygogonium* and *Cyanidium* rise in the morning, decrease during the afternoon, and show a second peak in the afternoon to evening (Fig. 7). This result is similar to that obtained for *Cyanidium* (Rothschild and Giver, 1993) and for a marine intertidal mat composed of the cyanobacterium *Lyngbya* (Rothschild and Giver, unpublished observations). The cause of this dip is unknown. A possible explanation is that the dip is caused by

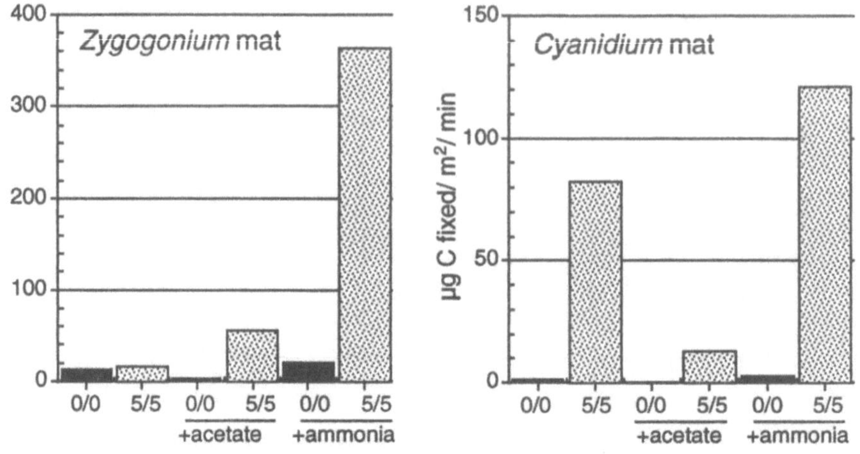

Figure 7. Effect of preincubation in elevated C_i on carbon fixation rates. Mat cores were incubated in stream water supplemented with 5 mM sodium bicarbonate before carbon fixation experiments, as described in the text. In each sample, the first number indicates the concentration of supplemental bicarbonate in which the samples were preincubated, and the second number indicates the concentration of supplemental bicarbonate added to the sample during the incubation. Each bar represents the average of four independent replicates. Bars representing samples that were preincubated and incubated with no additional bicarbonate are shown in black, and those incubated and preincubated with 5 mM bicarbonate are speckled. In all cases, carbon fixation was dramatically enhanced by additional bicarbonate during incubation, and this effect was actually enhanced by preincubation. The *Cyanidium* experiments were begun 23 June 1993, and the *Zygogonium* experiments four days later.

ultraviolet radiation because UV fluxes are highest at midday and laboratory studies have demonstrated a down-regulation in DNA synthesis in response to UV (Ronai et al., 1990; Livneth et al., 1993). To test this hypothesis, we built boxes of UV-transparent and UV-opaque Plexiglas and placed them over the mats in June 1996. After several days, DNA synthesis rates were measured (Fig. 7). The afternoon dip disappeared, but there was an overall decrease in DNA synthesis rates in the absence of UV. This result was surprising because previous work has shown that UV radiation decreases growth rate in algae (Bothwell et al., 1993 and references therein). But DNA synthesis is not limited to cellular reproduction. One type of DNA damage repair, excision repair, involved the removal of the damaged bases or bases plus a flanking region of DNA. The DNA is resynthesized using the undamaged strand as a template (Sancar, 1994). Excision repair makes the greatest contribution to radiation resistance in microorganisms (Nasim and James, 1978). Our current working hypothesis is that the difference in DNA synthesis rates represents the synthesis of new DNA as the result of excision repair in response to UV damage. We are currently testing this hypothesis with further experiments.

It should be noted that the calculation of the rates of synthesis assumed that there were no diurnal changes in the specific activity of the precursor pool, that is α-labeled deoxynucleotide triphosphates. Previous work on phytoplankton (Winn and Karl, 1986) suggests that this pool varies little during the day, but my lab is currently verifying these results for microbial mats. Uptake of phosphate is light-induced in some algae (Cembella et al., 1979), and is also monitored for *Cyanidium* and *Zygogonium*. The initial data (Rothschild,

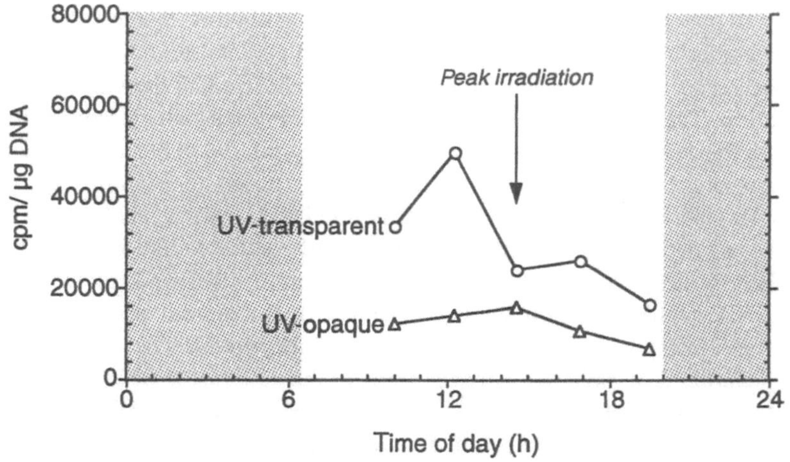

Figure 8. Diurnal patterns of *Cyanidium* DNA synthesis in the presence and absence of UV radiation. The experiments were conducted on 24 June 1993 by Charles Cockell. The shaded areas of the graph indicate the times of day when solar irradiance, as measured by photosynthetic photon flux, was not detected.

unpublished) suggest some light induction and effect of UV on phosphate uptake in *Cyanidium*, but variation among UV treatments appears trivial by late afternoon, when the biggest differences in DNA synthesis are detected. This suggests that at least the afternoon and evening results are not an artifact of differences in phosphate uptake but are indeed indicative of differences in DNA synthesis.

3.3. Partitioning of Photosynthate into DNA

Experiments were also conducted to determine the partitioning of photosynthate into DNA during the course of the day. Figure 8 shows the results of carbon fixation and the partitioning of photosynthate into DNA from experiments conducted 24 June 1993. The results show that a lower percentage of fixed carbon is partitioned into DNA after 1500 h, suggesting that DNA synthesis is a lower priority recipient of newly fixed carbon in the late afternoon. The partitioning pattern contrasts with total DNA synthesis patterns assessed by the incorporation of exogenously supplied [^{33}P]phosphate, as in Fig. 7. This result may indicate that DNA synthesis rates are not tightly coupled to carbon fixation rates through a constant partitioning of some portion of the fixed carbon to DNA.

4. DISCUSSION

4.1. Is There an Effect of High Temperature, Low pH, and Low pCO_2?

Acidophilic organisms such as *Cyanidium* and *Zygogonium* must cope with what to most organisms are extreme environmental parameters of low pH, low levels of C_i, and, in the case of *Cyanidium*, relatively high temperatures. Yet, the results presented here show that the overall pattern of carbon fixation is similar to that of other algal mats, plankton, and macroalgae studied. Temperature has been invoked to explain diurnal patterns of carbon fixation, and higher temperatures at midday are hypothesized as the cause of the dip in carbon fixation. The results presented for the *Cyanidium* mat, where temperature varies little through the day, argue against this explanation. Similarly, pH cannot explain diurnal patterns (as it might in drying tide pools) because pH remains constant through the day in these mats. Instead, carbon fixation is primarily regulated by the availability of radiation. Much of the regulation comes from photosynthetically active radiation (400 to 700 nm), although there is also some regulation by UV radiation (Cockell and Rothschild, submitted). For DNA synthesis, again there is a similarity in diurnal patterns between the acidophilic and intertidal mats (Cockell and Rothschild, 1996). The results suggest that radiation, including UV radiation, controls some of the patterns of DNA synthesis seen in nature, but other contributing factors have not been carefully explored.

Organisms that live in very acidic conditions are more limited in C_i than at more alkaline conditions, and thus respond dramatically to additional C_i. Yet, all of the C_i available to an organism in an acidic environment is virtually in the form of free CO_2. For comparison, the percentage DIC as free CO_2 in the ocean is ~0.7% at 18.5°C (Rothschild, 1994a). Because not all photosynthetic organisms (e.g., diatoms) have a mechanism to take up bicarbonate ions, such an acidic environment as Nymph Creek may not be quite as substrate-limiting as the total concentration of DIC suggests. The results presented in Fig. 4 show similar rates of carbon fixation on an areal basis in a hypersaline mat and an order of magnitude lower than an intertidal cyanobacterial mat (Rothschild, 1994b).

4.2. Ecological and Evolutionary Implications

High temperature, low pH, and low ambient levels of C_i may not control diurnal patterns of carbon fixation or DNA synthesis, but these parameters certainly affect the ecology of the ecosystem. At the most basic level, many organisms simply cannot survive in such conditions. For example, when the pH is less than 5 and the temperature is greater than 40°C, *Cyanidium* is the sole photosynthetic organism present (Doemel and Brock, 1971). Laboratory experiments show that *Cyanidium* can grow at a pH from near 0 to almost 5, and in a temperature range from 25 to 57°C (Brock, 1978). Though the pH optimum is broad ranging from pH just below 1 to 4, the temperature optimum is 45°C. The natural range is similar; *Cyanidium* is isolated from habitats that have a pH range of 0.9 to almost 5 and a temperature range of 13 to 57°C (Doemel and Brock, 1971). Although *Zygogonium* cannot tolerate as high a temperature as *Cyanidium* (e.g., the temperature optimum for photosynthesis is 25°C), it also has a maximal photosynthetic rate at low pH and a broad peak between pH 1 and 5 (Brock, 1978). Few other organisms, especially photosynthetic ones, can tolerate, much less thrive, under these environmental conditions.

Acidic communities have only a few organismal components so ecological interactions are restricted. For example, Belly et al. (1973) showed that the biogeochemical cycling of carbon in Nymph Creek is extremely simple. Field samples were incubated in the presence of $NaH^{14}CO_3$, and 2–4% of the fixed carbon was excreted as ^{14}C-organic material during a 4-hour incubation. The exact percentage varied with light level (Belly et al., 1973), so would presumably vary during the day. Belly et al. (1973) postulated that both the bacteria and fungal components of the mat use the excreted carbon. Fungal filaments increased in length in direct relationship to increasing numbers of *C. caldarium* cells/unit area; however, bacteria did not. This was interpreted to suggest that the fungi depend more directly on the excreted photosynthate than the bacteria whose growth rate must also depend on other factors such as the availability of inorganic nutrients. Belly et al. (1973) noted that the bacteria also produce CO_2 through respiration which the *Cyanidium* could then utilize in carbon fixation, whereas the *Cyanidium* in turn produce oxygen as a by-product of photosynthesis which is then available to the bacteria. Because CO_2 is so limiting in acidic hot springs, the presence of a population of actively respiring organisms may be important to the success of the alga.

An important ecological consequence of low pH is that C_i species, and therefore levels, are altered (Fig. 9) (e.g., Paasche, 1964). The following equations describe this relationship:

$$CO_2 + H_2O \leftrightarrow H^+ + HCO_3^- \qquad (2)$$

$$HCO_3^- \leftrightarrow H^+ + CO_3^{2-} \qquad (3)$$

where the pK_a of Eq. (2) is 6.37, and the pK_a of Eq. (3) is 10.25. At low pH, most of the C_i is in the form of CO_2, which is readily vented to the atmosphere resulting in low levels of DIC. High temperature increases the venting rate. The form of C_i available is important because the substrate for the carbon fixation enzyme ribulose-1,5-bisphosphate/carboxylase (RuBisCO) is CO_2. Even if HCO_3^- is transported across the cell membrane, it must be dehydrated before fixation. Not all organisms can take up bicarbonate, and thus from the point of view of carbon fixation, a low pH environment should be less restrictive. However, the relative energy costs of photosynthesis at a low pH and a low DIC are not clear. As pointed out by Raven and Lucas (1985), HCO_3^- uptake is more expensive energetically than CO_2 uptake. The energy used is in the form of ATP, which is photosynthetically produced. But, as the ratio of C_i to O_2 presented to RuBisCO decreases, there will be an increase in the ratio

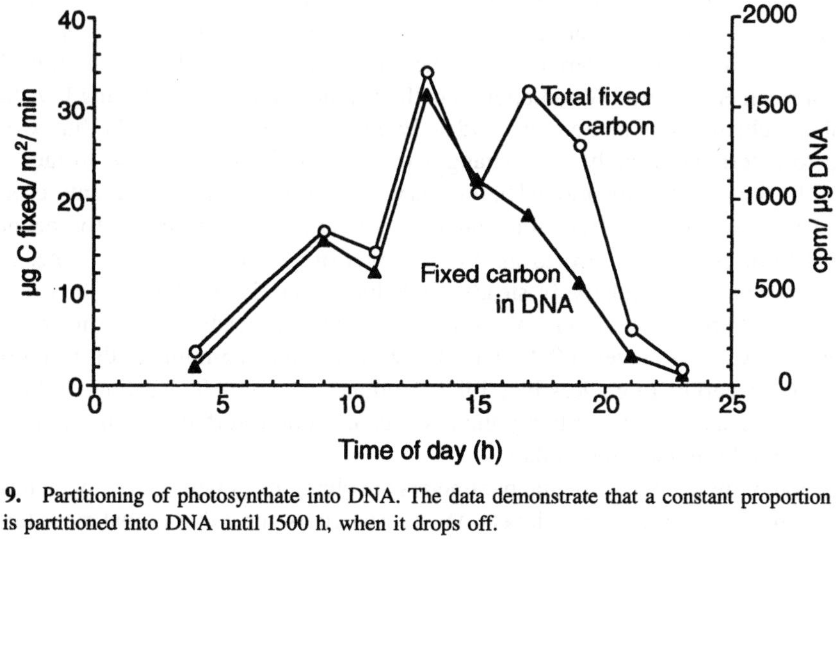

Figure 9. Partitioning of photosynthate into DNA. The data demonstrate that a constant proportion of fixed carbon is partitioned into DNA until 1500 h, when it drops off.

of oxygenase activity of the enzyme relative to the carboxylase activity, and the resulting photorespiration is expensive. Thus, a DIC accumulation mechanism should be advantageous in a low carbon, high light environment, and the ultimate energetic cost of living at low pH may be higher than living at higher pH.

The microbial communities in the acid hot springs of Yellowstone are, in several respects, excellent model systems to study early earth. The earliest forms were microbial, and microbes continued to dominate the earth's biota through much of the Precambrian. Environmentally, the early earth probably had features in common with the streams studied here. For example, global temperatures may have been high during the early evolution of life (e.g., Schwartzman et al., 1993). Levels of atmospheric CO_2, it is thought, were substantially higher during the Precambrian, and although they have fluctuated during the Phanerozoic, the changes have been minor in comparison (Kasting, 1993). Higher levels of atmospheric CO_2 could have caused higher levels of DIC in the water, especially surface water, although by how much is not known (Holland, 1984, p. 422). Holland and Kasting (1992) suggest that the concentration of HCO_3^- in Archaean seawater could have been 35–140 mM, substantially higher than the ~2 mM average of today's oceans. Carbon fixation rates were enhanced in mats exposed to elevated DIC. The partial pressure of CO_2 affects pH (e.g., Grotzinger and Kasting, 1993). With a higher level of atmospheric CO_2, aquatic ecosystems could have had a much lower pH because CO_2 forms an acid when it dissolves in water (Eq. (2)). The resulting acid could have been neutralized by immediate reaction with surface minerals (Schlesinger, 1991, p. 21) although there are competing models for Precambrian oceanic pH. Estimates range from as low as pH 6 during the early Archaean (Grotzinger and Kasting, 1993) to alkaline values of pH 10.5 at 3.5 Ga, 10 at 2 Ga, and dropping to modern levels by the end of the Precambrian (Kempe and Degens, 1985), although Holland (1992) believes that the pH probably rose during the Proterozoic in response to decreasing atmospheric levels of CO_2. Locally, pH can vary with depth in a microbial mat and in the same layer with time of day (Revsbech et al., 1983; Revsbech and Ward, 1984). Thus, the low pH of the streams studied may or may not provide models for early biotopes.

Surficial fluxes of UVB radiation were probably substantially higher during the Archaean. Solar luminosity has increased almost linearly through time from ~71% of current levels at 4.6 Ga to modern levels by the Cambrian (Gough, 1981; Kasting, 1987), but young stars can emit more UV radiation than the sun does today thus altering UV flux and atmospheric chemistry (Canuto et al., 1982). Even the average surface UV flux would be difficult to determine directly from changes in solar luminosity because surface flux is regulated by the variables mentioned before, as well as atmospheric constituents, especially ozone. There is much evidence that atmospheric O_2, the prerequisite for ozone, has increased from essentially zero to its present level of 21% over geologic time. Models suggest that pO_2 was a relatively steady 10^{-13} bars from 3.5 to 2.5 Ga, then increased rapidly to ~10^{-4} bars by 2.4 Ga, followed by a relatively steady increase to the present atmospheric level at sea level of 0.212 bar (0.209 atm) by the beginning of the Cambrian, or possibly much earlier (e.g., Kasting, 1987; Holland et al., 1989; Kasting, 1993). Thus, during the Precambrian and certainly the Archaean, average surface UV fluxes may have been substantially higher than today.

Several of these environmental parameters are changing, albeit relatively modestly, in the direction of their Precambrian levels. Most researchers agree that global mean temperature is increasing, atmospheric pCO_2 levels are rising (Boden et al., 1994), and this may be affecting aquatic levels of DIC (e.g., Jones and Levy, 1981). Surficial fluxes of UVB are rising as a result of the depletion of stratospheric ozone (Lloyd, 1993). The associated phenomena of increasing temperature and decreasing pH are becoming a problem for some temperate aquatic ecosystems.

5. SUMMARY

In conclusion, the following scenario is proposed. Low pH and high temperature, per se, do not influence diurnal patterns of carbon fixation or DNA synthesis, so temperature differences in the past or future will probably not alter these diurnal patterns. Absolute rates of carbon fixation are lower at low pH because DIC is severely limiting, and high temperature accentuates the problem. Under higher levels of DIC, carbon fixation rates will probably be higher. However, if this is accompanied by substantially higher pH, the percentage DIC as free CO_2 will decrease. If an organism does not have a mechanism to take up bicarbonate, the percentage usable DIC decreases. If the organism does have a bicarbonate uptake mechanism, carbon fixation rates may decrease because such mechanisms require energy that will then be diverted from other activities. DNA synthesis is regulated by other factors, including UV radiation. Increased surficial fluxes of UV radiation may well alter organismal timing and the amount of DNA synthesis. Reproductive rates remain difficult to estimate in microbial mats *in situ*. We had hoped that DNA synthesis could be used as a proxy for growth rates, but the data on the effect of UV radiation on DNA synthesis rates suggests that the relationship is not simple in nature.

ACKNOWLEDGMENTS. I would like to acknowledge, in the order they worked on this project, the following for technical assistance with the Yellowstone research and intellectual input: Rocco Mancinelli, Lori Giver, Claire Woodman Waite, Ben Squire, Charles Cockell, Genie Moore, and Lisa White, and Nat Pearson for editorial assistance.

REFERENCES

Alston, R. E. 1958. An investigation of the purple vacuolar pigment of *Zygogonium ericetorum* and the status of "algal anthocyanins" and "phycoporphyrins." *Am J. Bot.* **45:**688–692.

Bebout, B. M., Paerl, H. W., Crocker, K. M., and Prufert, L. E. 1987. Diel interactions of oxygenic photosynthesis and N_2 fixation (acetylene reduction) in a marine microbial mat community. *Appl. Environ. Microbiol.* **53:**2353–2362.

Belly, R. T., Tansey, M. R., and Brock, T. D. 1973. Algal excretion of ^{14}C-labeled compounds and microbial interactions in *Cyanidium caldarium* mats. *J. Phycol.* **9:**123–127.

Boden, T. A., Kaiser, D. P., Sepanski, R. J., and Stoss, F. W. (eds.). 1994. Trends '93: A compendium of data on global change. Publication No. ORNL/CDIAC-65, Carbon Dioxide Information Analysis Center, Oak Ridge National Laboratory, Oak Ridge, TN.

Bothwell, M. L., Sherbot, D., Roberge, A. C., and Daley, R. J. 1993. Influence of natural ultraviolet radiation on lotic periphytic diatom community growth, biomass accrual, and species composition: Short-term versus long-term effects. *J. Phycol.* **29:**24–35.

Brittain, A. M., and Karl, D. M. 1990. Catabolism of tritiated thymidine by aquatic microbial communities and incorporation of tritium into RNA and protein. *Appl. Environ. Microbiol.* **56:**1245–1254.

Brock, T. D. 1978. *Thermophilic microorganisms and life at high temperatures*. New York: Springer-Verlag.

Buch, K. 1960. Kohlendioxyd im Meerwasser. In Ruhland, W. (ed.), *Handbuch der Pflanzenphysiologie* (5: pp. 47–61). Berlin: Springer-Verlag.

Butterfield, N. J., Knoll, A. H., and Swett, K. 1990. A bangiophyte red alga from the Proterozoic of Arctic Canada. *Science* **250**:104–107.

Canuto, V. M., Levine, J. S., Augustsson, T. R., and Imhoff, C. L. 1982. UV radiation from the young sun and oxygen and ozone levels in the prebiological paleoatmosphere. *Nature* **296**:816–820.

Carman, K. R. 1990. Radioactive labeling of a natural assemblage of marine sedimentary bacteria and microalgae for trophic studies: An autoradiographic study. *Microb. Ecol.* **19**:279–290.

Cembella, A. D., Antia, N. J., and Harrison, P. J. 1979. The utilization of inorganic and organic phosphorous compounds as nutrients by eukaryotic microalgae: A multidisciplinary perspective. *CRC Crit. Rev. Microbiol.* **10**:317–391.

Cockell, C. S., and Rothschild, L. J. 1996. DNA synthesis patterns in algal mats. *J. Phycol.* **32**:12.

Cockell, C. S., and Rothschild, L. J. 1999. The effects of ultraviolet radiation A and B on diurnal variation in photosynthesis in three taxonomicell and ecologically diverse microbial mats. *Photochem. Photobiol.* **69**: 203–210.

Doemel, W. N., and Brock, T. D. 1971. The physiological ecology of *Cyanidium caldarium*. *J. Gen. Microbiol.* **67**:17–32.

Doyle, J., and Doyle, J. 1990. Isolation of plant DNA from fresh tissue. *Focus* **12**:13–15.

Fuhrman, J. A., and Azam, F. 1980. Bacterioplankton secondary production estimates for coastal waters of British Columbia, Antarctica, and California. *Appl. Environ. Microbiol.* **39**:1085–1095.

Gerasimenko, L. M., and Miller, Y. M. 1985. Gaseous exchange of the extreme thermophilic acidophilic phototrophic organism *Cyanidium caldarium*. *Mikrobiologiya* **54**:693–698.

Gilmour, C. C., Leavitt, M. E., and Shiaris, M. P. 1990. Evidence against incorporation of exogenous thymidine by sulfate-reducing bacteria. *Limnol. Oceanogr.* **35**:1401–1409.

Gorham, E. 1996. Lakes under a three-pronged attack. *Nature* **381**:109.

Gough, D. O. 1981. Solar interior structure and luminosity variations. *Solar Phys.* **74**:21–34.

Gromov, B. V., Avilov, I. A., Andreev, L. V., Lyutikov, V. N., and Nilitina, V. N. 1979. Acidophilic algae *Cyanidium caldarium* from Far Eastern USSR thermal springs. *Biol. Nauti (Mosc.)* **10**:78–82.

Grotzinger, J. P., and Kasting, J. F. 1993. New constraints on Precambrian ocean composition. *J. Geol.* **101**: 235–243.

Han, T. M., and Runnegar, B. 1992. Megascopic eukaryotic algae from the 2.1-billion-year-old Negaunee iron-formation, Michigan. *Science* **257**:232–235.

Hoffman, W. E., and Dawes, C. J. 1980. Photosynthetic rates and primary production by two Florida benthic red algal species from a salt marsh and a mangrove community. *Bull. Marine Sci.* **30**:358–364.

Holland, H. D. 1984. *The chemical evolution of the atmosphere and oceans*. Princeton, NJ: Princeton University Press.

Holland, H. D. 1992. Chemistry and evolution of the Proterozoic Ocean. In Schopf, J. W., and Klein, C. (eds.), *The Proterozoic biosphere* (pp. 169–172). Cambridge: Cambridge University Press.

Holland, H. D., Feakeo, C. R., and Zbinden, E. A. 1989. The flin flon paleosol and the composition of the atmosphere 1.8 BYBP. *Am. J. Sci.* **289**:362–389.

Holland, H. D., and Kasting, J. F. 1992. The environment of the Archean earth. In Schopf, J. W., and Klein, C. (eds.), *The Proterozoic biosphere* (pp. 21–24). Cambridge: Cambridge University Press.

Jeffrey, W. H., Paul, J. H., Cazares, L. H., DeFlaun, M. F., and David, A. W. 1990. Correlation of nonspecific macromolecular labelling with environmental parameters during [3H]thymidine incorporation in the waters of southwest Florida. *Microb. Ecol.* **20**:21–35.

Jones, E. P., and Levy, E. M. 1981. Oceanic CO_2 increase in Baffin Bay. *J. Marine Res.* **39**:405–416.

Kasting, J. F. 1987. Theoretical constraints on oxygen and carbon dioxide concentrations in the Precambrian atmosphere. *Precambr. Res.* **34**:205–229.

Kasting, J. F. 1993. Earth's early atmosphere. *Science* **259**:920–926.

Kempe, S., and Degens, E. T. 1985. An early soda ocean? *Chem. Geol.* **53**:95–108.

Knoll, A. H. 1992. The early evolution of eukaryotes: A geological perspective. *Science* **256**:622–627.

Livneh, Z., Cohen, O., Skaliter, R., and Elizur, T. 1993. Replication of damaged DNA and the molecular mechanism of ultraviolet light mutagenesis. *Crit. Rev. Biochem. Mol. Biol.* **28**:465–513.

Lloyd, S. A. 1993. Stratospheric ozone depletion. *Lancet* **342**:1156–1158.

Long, S. P., and Drake, B. G. 1992. Photosynthetic CO_2 assimilation and rising atmospheric CO_2 concentrations.

In Baker, N. R., and Thomas, H. (eds.), *Crop photosynthesis: Spatial and temporal determinants* (pp. 69–103). B. V., Amsterdam: Elsevier Science.

Nasim, A., and James, A. P. 1978. Life under conditions of high irradiation. In Kushner, D. J. (ed.), *Microbial life in extreme environments* (pp. 409–439). New York: Academic Press.

Oechel, W. C., Cowles, S., Grulke, N., Hastings, S. J., Lawrence, B., Prudhomme, T., Riechers, G., Strain, B., Tissue, D., and Vourlitis, G. 1994. Transient nature of CO_2 fertilization in Arctic tundra. *Nature* **371:**500–503.

Ott, F. D., and Seckbach, J. 1994. A review on the taxonomic position of the algal genus *Cyanidium* Geitler 1933 and its ecological cohorts *Galdieria* Merola in Merola et al. (1981) and *Cyanidioschyzon* De Luca, Taddei and Varano (1978). In Seckbach, J. (ed.), *Evolutionary pathways and enigmatic algae: Cyanidium caldarium (Rhodophyta) and related cells* (pp. 113–132). Dordrecht, The Netherlands: Kluwer Academic.

Paasche, E. 1964. A tracer study of the inorganic carbon uptake during coccolith formation and photosynthesis in the coccolithophorid *Coccolithus huxleyi. Physiol. Plant. Supplement* **3:**1–82.

Peterson, B. J. 1980. Aquatic primary productivity and the ^{14}C-CO_2 method: A history of the productivity problem. *Annu. Rev. Ecol. Syst.* **11:**359–385.

Peterson, R. B., Friberg, E. E., and Burris, R. H. 1977. Diurnal variation in N_2 fixation and photosynthesis by aquatic blue-green algae. *Plant Physiol.* **59:**74–80.

Powles, S. B. 1984. Photoinhibition of photosynthesis induced by visible light. *Annu. Rev. Plant Physiol.* **35:** 15–44.

Ragan, M. A., and Gutell, R. R. 1995. Are red algae plants? *Bot. J. Linn. Soc.* **118:**81–105.

Raven, J. A., and Lucas, W. J. 1985. Energy costs of carbon acquisition. In Lucas, W. J., and Berry, J. A. (eds.), *Inorganic carbon uptake by aquatic photosynthetic organisms* (pp. 305–324). Rockville, MD: American Society of Plant Physiologists.

Revsbech, N. P., and Ward, D. M. 1984. Microelectrode studies of interstitial water chemistry and photosynthetic activity in a hot spring microbial mat. *Appl. Environ. Microbiol.* **48:**270–275.

Revsbech, N. P., Jørgensen, B. B., Blackburn, T. H., and Cohen, Y. 1983. Microelectrode studies of the photosynthesis and O_2, H_2S, and pH profiles of a microbial mat. *Limnol. Oceanogr.* **28:**1062–1074.

Ronai, Z. A., Lambert, M. E., and Weinstein, I. B. 1990. Inducible cellular responses to ultraviolet light irradiation and other mediators of DNA damage in mammalian cells. *Cell. Biol. Toxicol.* **6:**105–126.

Rothschild, L. J. 1991. A model for diurnal patterns of carbon fixation in a Precambrian microbial mat based on a modern analog. *BioSystems* **25:**13–23.

Rothschild, L. J. 1994a. CO_2 and diatom mats. *Nature* **368:**817.

Rothschild, L. J. 1994b. Elevated CO_2: Impact on diurnal patterns of photosynthesis in natural microbial ecosystems. *Adv. Space Res.* **14:**285–289.

Rothschild, L. J., and Giver, L. J. 1993. Diurnal patterns of carbon fixation and DNA synthesis in a *Cyanidium caldarium* mat community. *J. Euk. Microbiol.* **40:**14A.

Rothschild, L. J., and Heywood, P. 1988. "Protistan" nomenclature: Analysis and refutation of some potential objections. *BioSystems* **21:**197–202.

Rothschild, L. J., and Mancinelli, R. L. 1990. A model for the evolution of carbon fixation in microbial mats, 3500 million years ago to the present. *Nature* **345:**710–712.

Sancar, A. 1994. Mechanisms of DNA excision repair. *Science* **266:**1954–1956.

Schlesinger, W. H. 1991. *Biogeochemistry, an analysis of global change.* San Diego, CA: Academic Press.

Schwartzman, D., McMenamin, M., and Volk, T. 1993. Did surface temperatures constrain microbial evolution? *Bioscience* **43:**390–393.

Seckbach, J., and Ott, F. D. 1994. Systematic position and phylogenetic status of *Cyanidium* Geitler 1933. In Seckbach, J. (ed.), *Evolutionary pathways and enigmatic algae: Cyanidium caldarium (Rhodophyta) and related cells* (pp. 133–143). Dordrecht, The Netherlands: Kluwer Academic.

Staley, J. T., and Konopka, A. 1985. Measurement of *in situ* activities of nonphotosynthetic microorganisms in aquatic and terrestrial habitats. *Annu. Rev. Microbiol.* **39:**321–346.

Strickland, J. D. H., and Parsons, T. R. 1968. Photosynthetic rate measurements. In: *A practical handbook of seawater analysis*, Bulletin 167 (pp. 261–278). Ottawa: Fisheries Research Board of Canada.

Thorn, P. M., and Ventullo, R. M. 1988. Measurement of bacterial growth rates in subsurface sediments using the incorporation of tritiated thymidine into DNA. *Microb. Ecol.* **16:**3–16.

Villbrandt, M., Stal, M. J., and Krumbein, W. E. 1990. Interactions between nitrogen fixation and oxygenic [sic] photosynthesis in a marine cyanobacterial mat. *FEMS Microbiol. Lett.* **74:**59–72.

Vincent, W. F. 1980. Mechanisms of rapid photosynthetic adaptation in natural phytoplankton communities. II.

Changes in photochemical capacity as measured by DCMU-induced chlorophyll fluorescence. *J. Phycol.* **16:**568–577.

Walter, M. R., Bauld, J., Des Marais, D. J., and Schopf, J. W. 1992. A general comparison of microbial mats and microbial stromatolites: Bridging the gap between the modern and the fossil. In Schopf, J. W., and Klein, C. (eds.), *The Proterozoic biosphere* (pp. 335–338). Cambridge: Cambridge University Press.

Warburg, O., and Christian, W. 1942. Isolierung und kristallisation des gärungsferments enolase. *Biochem. Zeit.* **310:**384–421.

Ward, D. M., Weller, R., Shiea, J., Castenholz, R. W., and Cohen, Y. 1989. Hot spring microbial mats: Anoxygenic and oxygenic mats of possible evolutionary significance. In Cohen, Y., and Rosenberg, E. (eds.), *Microbial mats. Physiological ecology of benthic microbial communities* (pp. 3–15). Washington, D.C.: American Society for Microbiology.

Wellsbury, P., Herbert, R. A., and Parkes, R. J. 1993. Incorporation of [methyl-^{3}H]thymidine by obligate and facultative anaerobic bacteria when grown under defined culture conditions. *FEMS Microbiol. Lett.* **12:** 87–95.

Winn, C. D., and Karl, D. M. 1986. Diel nucleic acid synthesis and particular DNA concentrations: Conflicts with division rate estimates by DNA accumulation. *Limnol. Oceanogr.* **31:**637–645.

Yan, N. D., Keller, W., Scully, N. M., Lean, D. R. S., and Dillon, P. J. 1996. Increased UV-B penetration in a lake owing to drought-induced acidification. *Nature* **381:**141–143.

11

The Zonation and Structuring of Siliceous Sinter around Hot Springs, Yellowstone National Park, and the Role of Thermophilic Bacteria in Its Deposition

Donald R. Lowe, Kai S. Anderson, and Deena Braunstein

1. INTRODUCTION

Studies of early terrestrial life have shown that before about 2 billion years ago (Ga) the Earth's biota was dominated by prokaryotic bacteria, possibly including photosynthetic cyanobacteria. As a geologic record, these early organisms have left carbonaceous microfossils, inferred to be the remains of individual filamentous and coccoidal bacteria, and flat, domal, and columnar laminated structures called stromatolites, which are interpreted to have formed through the vertical accretion of bacterial mat communities (Walter, 1976a). Stromatolites are abundant in thick Precambrian carbonate formations that range in age from about 3.0 Ga to the end of the Precambrian, about 560–530 million years ago (Ma). Precambrian carbonaceous microfossils are best preserved in siliceous rocks, called cherts, that are composed largely of microcrystalline quartz. The organisms preserved in cherts were entombed through silicification before they could be destroyed by biological degradation or chemical attack. Cherts and related siliceous deposits are widely preserved in thick sedimentary and volcanic sequences that accumulated in the Archaean (>2.5 Ga) and Proterozoic (2.5–0.53 Ga) oceans.

Sedimentological and paleobiological studies of these ancient rock sequences inevit-

Donald R. Lowe, Kai S. Anderson, and Deena Braunstein • Department of Geological and Environmental Sciences, Stanford University, Stanford, California 94305-2115.

Thermophiles: Biodiversity, Ecology, and Evolution, edited by Reysenbach *et al.* Kluwer Academic/Plenum Publishers, New York, 2001.

ably raise questions concerning the relative importance of organisms versus abiotic processes in forming putative biogenic features. A controversy has long existed concerning the biogenicity of microscopic carbonaceous spheroids in early Precambrian rocks because such structures can form through both biological and nonbiological processes (e.g., Schopf, 1975, 1976, 1983). Recently, some of the most ancient structures interpreted as biological stromatolites (Walter et al., 1980; Lowe, 1980; Byerly et al., 1986) have been reinterpreted to have formed largely or entirely through abiotic processes (Lowe, 1994).

Discriminating between biotic and abiotic structures is particularly problematic in Precambrian deposits. The silica and carbonate budgets of modern oceans are dominated by bioprecipitation of organisms to form hard parts. Carbonate is deposited mainly by organisms living in shallow marine waters, where they have accumulated to form thick limestone formations characteristic of sedimentary sequences younger than about 2.6–2.5 billion years. Siliceous hard parts, principally opaline radiolarian tests, diatom frustules, and sponge spicules, accumulate after the deaths of the organisms and, with burial, are transformed through dissolution/precipitation and recrystallization into silica-rich deposits, including chert. All known silica-secreting organisms and most carbonate-secreting organisms evolved since the end of the Precambrian, about 530 million years ago, and it is generally thought that silica and carbonate budgets of the Precambrian oceans were regulated by inorganic precipitation and by interactions between seawater and surrounding rocks and mineral grains (Siever, 1992; Maliva et al., 1989).

Unique features formed through biological activities that are absent in similar abiogenic structures are essential to the identification of biological structures in the geologic record. The distinctive morphologies of shells and organism skeletal parts in Phanerozoic rocks make identifying them relatively easy. Such features are presently less well defined for most bacterial fossils, including stromatolites and simple carbonaceous microfossils, which comprise the biological record of the early Precambrian. Establishing the biogenicity of these structures will depend on studying both biological and abiological stromatolite- and microfossil-like structures and understanding their formative processes. Although bacterial stromatolites are less common in today's oceans than they were in Precambrian time due to the activities of grazing invertebrates, there are a number of localities, such as Shark Bay, Western Australia (Playford and Cockbain, 1976), and the Bahamas (Dill et al., 1986), where modern stromatolites can be studied. These are primarily carbonate-depositing systems, and they provide important information on the morphology and ecology of modern carbonate stromatolites that can be applied to understanding similar features preserved in the geologic record. Studies of both modern and ancient carbonate stromatolites (Hoffman, 1976; Grotzinger, personal communication) and abiotic carbonate stromatolite-like structures, such as cave formations (Thrailkill, 1976), indicate that stromatolite geometry and structuring is or can be controlled largely by nonbiological processes. The problem still remaining is to determine which features of stromatolite-like structures unequivocally indicate biological activities.

Most siliceous stromatolites in the geologic record were originally composed of carbonate that was replaced by silica after deposition. However, some of the most ancient stromatolite-like structures yet reported, 3.3 billion years old from the Barberton Greenstone Belt in South Africa (Byerly et al., 1986), have recently been reinterpreted as possibly having formed as nonbiogenic silica deposits (Lowe, 1994). It is possible that many laminated stromatolite-like deposits were formed in the Archaean oceans through abiotic silica deposition.

The principal modern settings where silica is being deposited through abiotic precipitation and also where bacterial mats are well developed and subject to early, rapid silicification are is around hot springs. Alkaline hot springs in many geothermal areas are surrounded by deposits of opaline silica. These deposits, termed siliceous sinter or simply sinter, form because the hot, subsurface geothermal waters contain high levels of dissolved silica leached from the volcanic rocks through which the geothermal waters flow. As the water erupts via geysers or hot springs and enters surficial outflow systems, it cools and the silica polymerizes, solubility drops as a function of temperature, and silica is deposited as sinter around the vent and down the outflow system (White et al., 1956). Organisms thrive in these geothermal waters from the boiling springs to the surrounding low-temperature meadows and marshes. Bacteria predominate throughout much of the system, and from about 60°C to 30°C, extensive, commonly silicified bacterial mats characterize most outflow systems (Walter et al., 1976; Brock and Starr, 1978). In addition, much sinter formed in the high-temperature parts of hot springs and their outflow systems, termed geyserite (White et al., 1964; Walter, 1976b), commonly exhibits laminated columnar structures that closely resemble biological stromatolites (Walter, 1976b).

It is clear that ambiguities regarding the biological versus abiological origin of laminated structures in the Precambrian geologic record are equally valid in the case of siliceous sinter deposited around modern hot springs. Studies to date suggest that virtually every pool and outflow surface that is wet is populated by bacteria, ranging from isolated individual bacteria to submillimeter-thick biofilms to dense bacterial mats up to several centimeters thick. In natural settings, the absence of sterile surfaces makes it impossible to establish the morphology and structuring of silica deposited under strictly abiotic conditions and therefore to distinguish unambiguously between the biotic and abiotic controls on sinter structuring. Cady and Farmer (1996) and Jones et al. (1997) provide considerable evidence that bacteria influence the microstructure (<1 mm) of sinter deposited in splash zones around the rims of hot springs at temperatures near boiling. As will be discussed here, the variations in sinter morphology and structuring among the various pool and geyser types in Yellowstone suggest to us that macrostructural features (>1 cm) are controlled largely by pool hydrodynamics, including the eruption style, the amount of splash and surging, circulation patterns within the pools, and evaporation.

To better understand the origin, structuring, and potential biogenicity of Precambrian siliceous deposits and to help develop criteria for distinguishing between biogenic and abiogenic siliceous structures in the geologic record, we initiated a study of modern sinter around thermal springs in Yellowstone National Park in 1991. This study has focused to date on characterizing the principal depositional zones within the outflow systems, the main types of sinter (sinter facies), and the relationship between sinter deposition and structuring, chemical and physical settings, and bacterial mats. We hope that the study of thermophilic bacteria that inhabit sinter mounds in Yellowstone will help to resolve some of the controversies about the origin of possible biogenic structures in rocks deposited in the Earth's earliest oceans. This report presents preliminary results of that study.

2. GEOLOGIC SETTING OF YELLOWSTONE GEOTHERMAL SYSTEM

Yellowstone National Park is situated above the Yellowstone hot spot, a thermal anomaly that has been active within the western United States for at least the last 16 million years (Morgan, 1972; Suppe et al., 1975). During this time, the hot spot has moved generally toward the east northeast, from the area of southeastern Oregon 16 million years ago, across the area of the Snake River Plain, to the present site of active volcanism in Yellowstone National Park (Pierce and Morgan, 1992). This apparent eastward movement of the volcanic center reflects the westward movement of the North American Plate over the sublithospheric magma source. The hot spot is currently situated beneath the western part of the ancient Precambrian continental crust of North America. Melting has involved partial fusion of the continental crust, and consequently the eruptive magmas in Yellowstone are composed largely of highly siliceous volcanic rocks, mainly rhyolites (Christiansen, 1984).

During the last 2 million years, magmatic activity in the Yellowstone area has involved the formation of a series of large calderas (Christiansen, 1984). The latest formed about 600,000 years ago in the southwestern part of the present Yellowstone National Park (Fig. 1).

The bulk of modern geothermal activity in Yellowstone is centered within or around the margin of this most recent caldera (Fig. 1). Much of this activity, including that in the Upper and Lower Geyser Basins, the Shoshone Geyser Basin, and the West Thumb Geyser Basin, is alkaline and is accompanied by the deposition of siliceous sinter. Norris Geyser Basin, located just north of the caldera, is mainly an acidic system. At Mammoth Hot

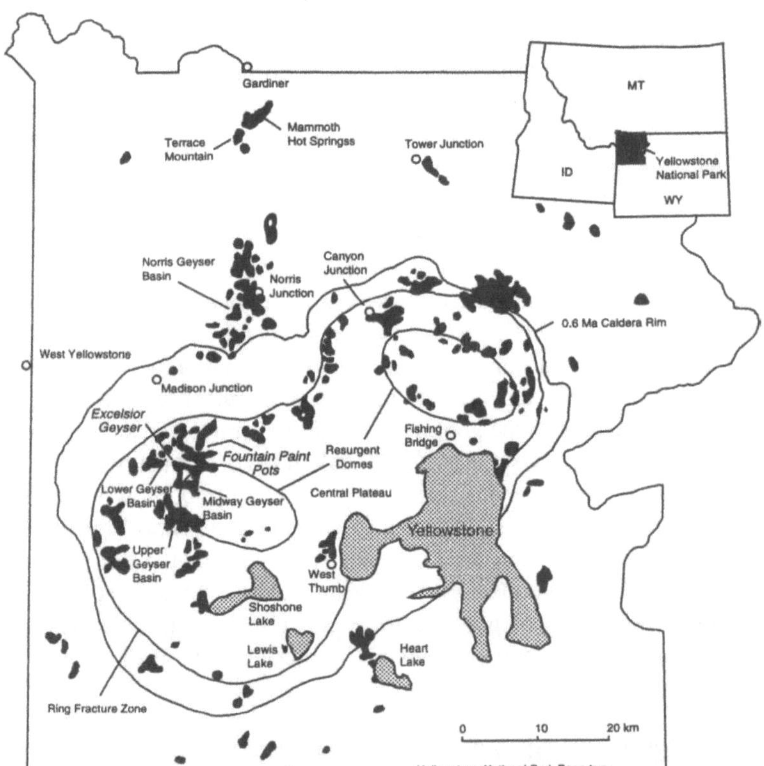

Figure 1. Map of the Yellowstone area showing the most recent caldera boundaries and the principal geothermal area.

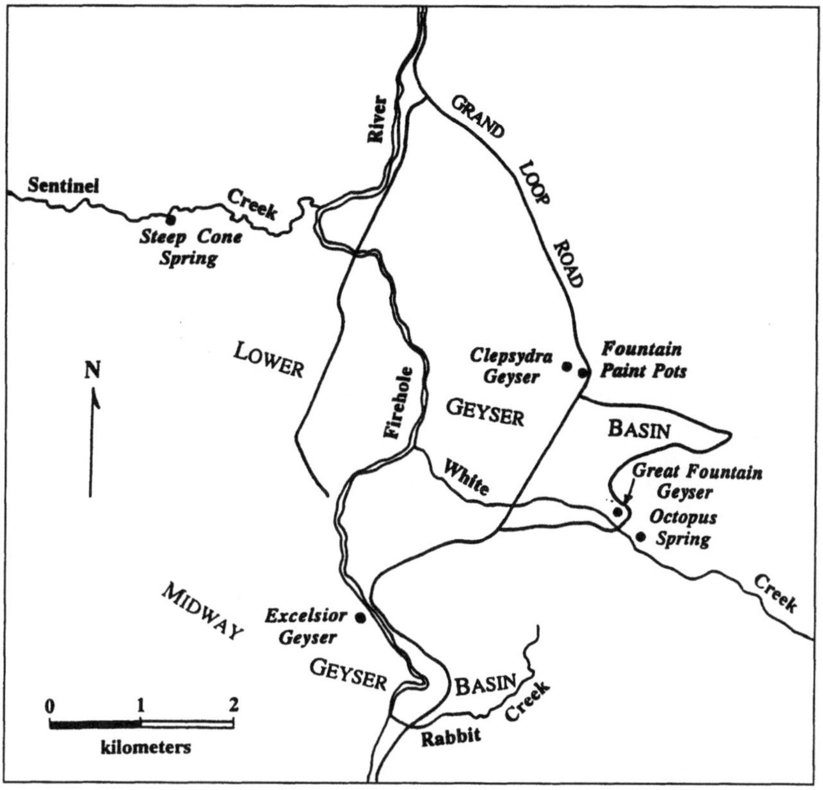

Figure 2. Map of the Lower Geyser Basin showing hot springs and geysers referred to in the text.

Springs, near the northern edge of the Park, geothermal waters leach carbonate from buried Paleozoic limestones and dolomites and, upon eruption, form calcareous deposits of travertine.

The present study has focused on three areas in Lower Geyser Basin: (1) a large sinter mound at Fountain Paint Pot; (2) Steep Cone Spring in Sentinel Meadows; and (3) old sinter deposits exposed in the wall of Excelsior Geyser (Fig. 2). In addition, studies of high-temperature sinter are underway by Braunstein in these areas and in Shoshone Geyser Basin. In the course of this research, we have collected samples of siliceous sinter from the outflow systems of numerous hot springs and geysers of all types throughout the Park and have cut, slabbed, and studied the sinter structuring and texture in hand specimens, slabs, and thin sections. Detailed maps of sinter and bacterial zonation have been drawn for the Fountain Group and Steep Cone Spring outflow systems (Anderson et al., unpublished). Stratigraphic sections have been measured through older sinter deposits at Excelsior Geyser.

3. MORPHOLOGICAL SUBDIVISIONS OF OUTFLOW SYSTEMS

Alkaline springs and geysers in Yellowstone are commonly located atop large, convex-upward deposits of siliceous sinter termed sinter mounds. The largest spring and

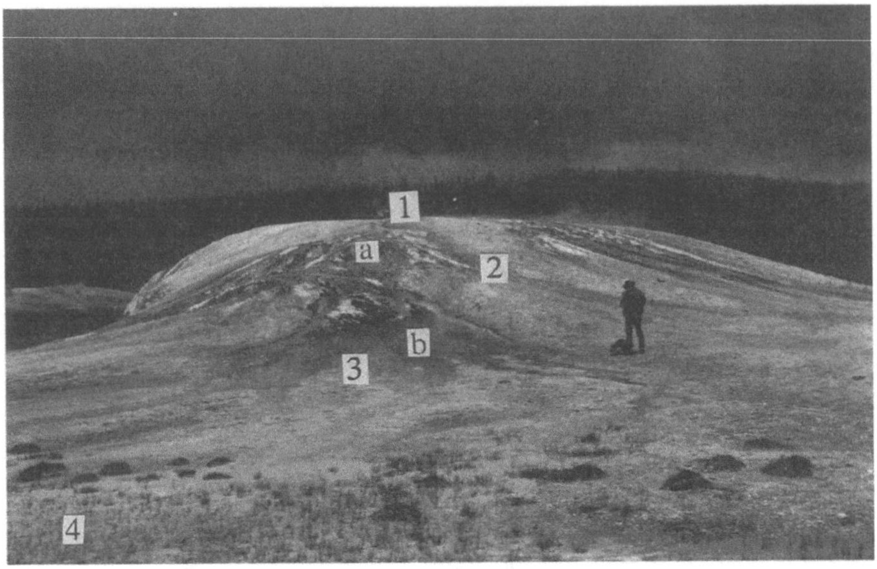

Figure 3. The sinter mound of Steep Cone Spring in Sentinel Meadow. The two visible active outflow systems are fed by the circular boiling Steep Cone Spring (Fig. 4) on the flat mound top. The outflow system on the left shows (1) the lower part of the proximal outflow channel, (2) the medial channel complex, (3) the distal channel-debris apron, and (4) the diatom marsh and meadow. Recognizable biological zones include (a) the *Phormidium* zone that consists of blotchy medium gray (living mats) and white (dead *Phormidium* sinter) areas and (b) the *Calothrix* zone, a uniformly dark area on the lower part of the wetted outflow system and flanking the *Phormidium* mats. A map of this system is shown in Fig. 5.

geyser groups have locally formed sinter mounds as much as 50 meters thick and a thousand meters across. One of the best examples of a small sinter mound is that of Steep Cone Spring (Fig. 3) in Sentinel Meadow. It is a roughly circular mound, approximately 5–7 meters high and 50–60 meters in diameter and has a small circular continuously boiling spring at the top (Fig. 4 and Walter, 1976b; Fig. 3). The mound has been cut into on the west side by Sentinel Creek, revealing the internal structuring of the siliceous deposits.

The outflow systems of most continuously flowing springs and geysers can be subdivided based on (1) the morphology of the sinter surface and (2) the water temperature and type of bacterial mats. These two means of subdivision are largely independent and will be considered separately.

We have recognized five main morphological zones in continuously active outflow systems: (1) the vent pool and near-vent outflow area, (2) the proximal outflow channel, (3) the channel and terrace complex, (4) the distal debris apron, and (5) the surrounding marsh and meadow (Figs. 3 and 5).

3.1. Vent Pool and Near-Vent Outflow Area

The vent and near-vent areas can be subdivided on the basis of outflow hydraulics. The principal subdivisions include (1) fully submerged sides and bottoms of pools, (2) intermit-

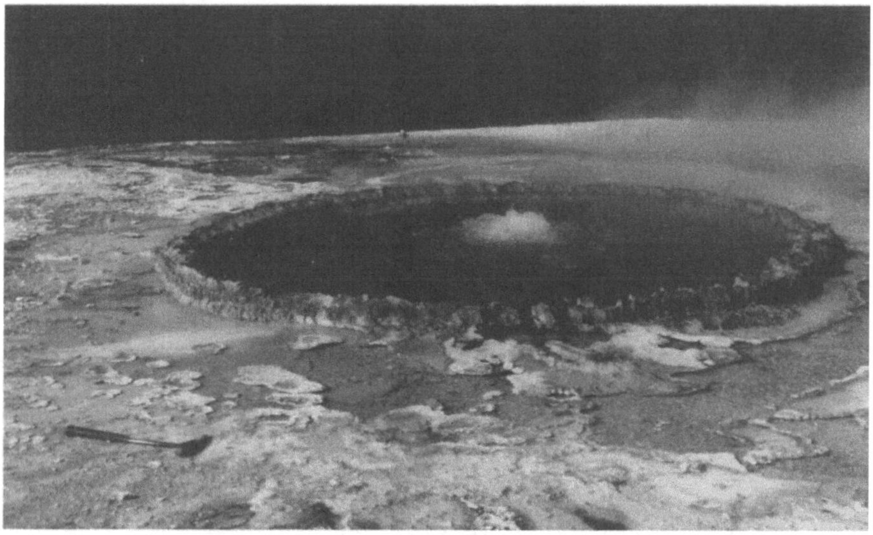

Figure 4. Steep Cone Spring. The boiling spring is enclosed by a continuous sinter rampart about 10–15 cm high breached by a single main channel about 5 cm wide near the center of the photograph. The sinter rampart is surrounded by a shallow, continuously filled near-vent outflow pond that feeds two outflow systems to the right (Fig. 3) and a third small system in the background. The very shallow near-vent outflow pond is characterized by water temperatures above 73°C and thin, flat, lily-pad-like sinter.

tently submerged spring margins, (3) splash zones around springs and geysers, and (4) near-vent outflow ponds and basins.

3.1.1. Fully Submerged Pool Sides and Bottoms

Many springs and geysers are characterized by more or less permanent pools (Figs. 4 and 6). The submerged sides and bottoms of these pools are commonly sites of sinter deposition. Where sinter is deposited subaqueously through direct precipitation, the sides and bottoms of pools consist of large botryoidal deposits made up of layers 1 to 4 cm thick, composed of tightly packed radiating prismatic or spicular sinter (Walter, 1976b, pp. 88–89). Long-term sinter deposition within vents can choke the inflow of subsurface waters, eventually inactivating the spring or geyser. The sides and bottoms of other pools accumulate fine, buff to tan, particulate siliceous deposits. Many springs show sides and subaqueous surfaces that are sites of nondeposition or even erosion, suggesting that one mechanism of spring development involves the dissolution or the physical quarrying and removal through explosive eruptions of older deposits.

3.1.2. Intermittently Submerged Spring Margins

Most hot spring pools show changes in water level that result in zones around their margins that are exposed and dry at times and are submerged and wet at others. These zones are analogous to the intertidal zone of marine systems. This facies is especially well-

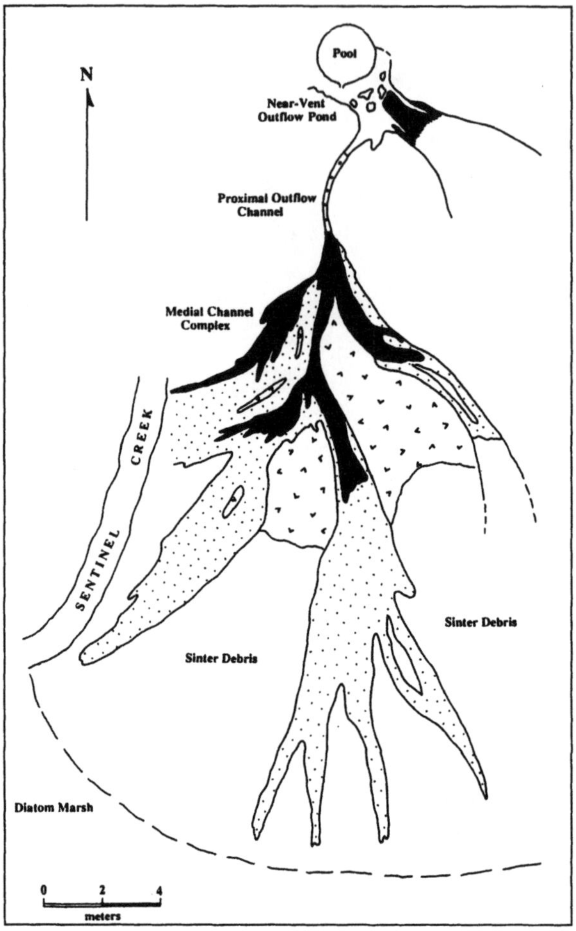

Figure 5. Map of the morphological and bacterial zones of the main outflow system on the sinter mound of Steep Cone Spring. The map was drawn in the summer of 1992. Bacterial zone symbols: large dots, *Synechococcus-Chloroflexus* zone; black, *Phormidium* zone; fine stipple, *Calothrix* zone. Small v's indicate inactive bacterial sinter.

developed around highly active systems, like boiling and violently surging springs (Fig. 4). It also characterizes the rims of placid surging springs, such as Octopus Spring, that show the quiescent rise and fall of water level, from less than a centimeter to several tens of centimeters during periods of a few minutes to hours. Very narrow intermittently submerged zones occur even around the rims of nonsurging springs due to disturbances of the water surface caused by the wind and boiling. Sinter within the intermittently submerged zone accumulates mainly through water evaporation from wetted surfaces.

Sinter deposited in intermittently submerged zones displays a variety of morphologies and structuring, largely related to the magnitude of fluctuation in water level and the rates of evaporation versus outflow. In relatively quiescent, nonsurging to gently surging pools, thin flat layers of sinter, from 0.2 to 2 cm thick, that have rounded lily-pad-like to elongate

Figure 6. Vent pool and large near-vent outflow basin at Tardy Geyser in Upper Geyser Basin. The large circular near-vent basin fills during eruptions, when it also temporarily lies within the splash zone. This pool does not flow continuously so there are no bacterial mats associated with the outflow system. The diameter is approximately 5 meters.

finger-like projections are deposited in a narrow zone around the rim and extend horizontally out at or slightly above the water surface (Fig. 7). Internally, this sinter consists of layers less than 0.5 mm thick that are progressively accreted onto the near-pool upper and outer surfaces, forming laminated structures closely resembling stromatolites that grow poolward (Fig. 8).

3.1.3. Splash Zones

Geysers, highly active surging springs, and boiling springs periodically or continuously produce airborne water masses, droplets, and mist. These fall in close proximity to the vents, run off, and leave surfaces to dry or partially dry between wettings. In many springs, there is a significant degree of overlap between intermittently submerged spring margins and splash zones (Figs. 4 and 6). Sinter deposited in these zones is some of the most varied and complex observed on sinter mounds (see Walter, 1976b, Figs. 4, 5, 10–13). Much of this sinter has a columnar, mound-like, or spicular form (Walter, 1976b, Figs. 10–12), where active growth sites are separated by depressions, grooves, or low areas where sinter deposition occurs less rapidly. Well-defined rims, ramparts, and levee-like sinter buildups characterize the splash zones of boiling pools (Fig. 4). The surfaces of splash zone sinter deposits tend to be hard because small loose particles are quickly removed by the falling water. The sinter commonly has a laminated structure that produces internally laminated columns, spicules, and sheets that closely resemble ancient stromatolites (Fig. 9

Figure 7. Octopus Spring. Photo shows a part of the main pool (upper right) with light bottom lacking visible bacterial mats and water temperatures above 73°C, discontinuous pool rim made up of lily-pad-like to finger-like sinter, and marginal near-vent outflow pond (left) with bottom covered by dark *Synechococcus-Chloroflexus* mats.

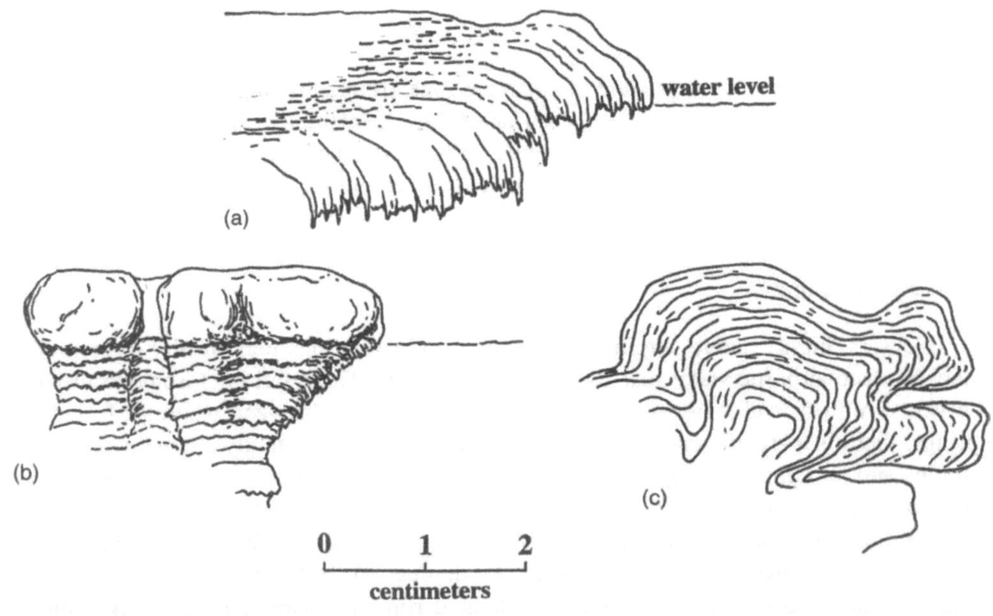

Figure 8. Sketches of finger-like sinter developed along margins of high-temperature, quiescent to gently surging pools like Octopus Spring (Fig. 7). (a) cross section showing lateral growth by addition of sinter mainly along the outer wet edge at and just above the water level combined with upward growth that accompanies rising pool water level. (b) general sketch of lateral terminations of digits in sinter shown in (a). (c) plan view of underside of finger-like sinter in (b) showing fine growth laminations and resemblance to biological stromatolites.

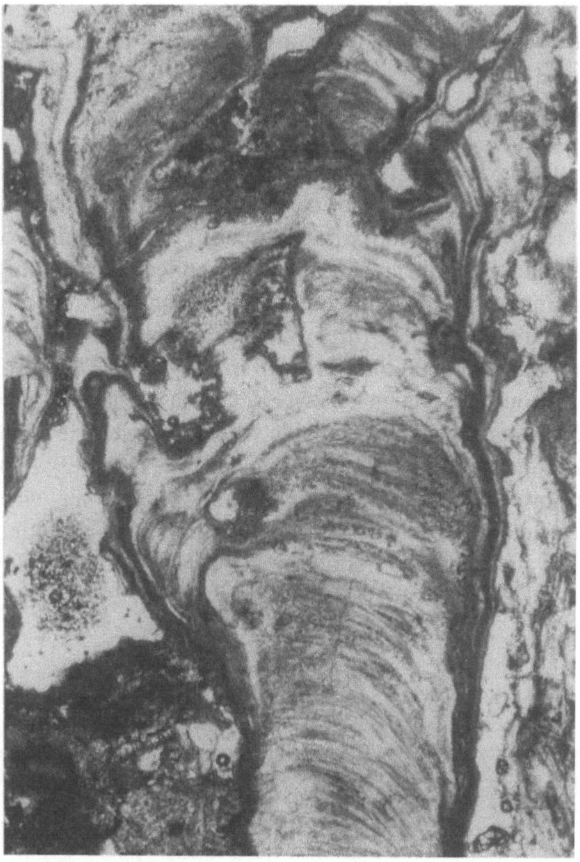

Figure 9. Photomicrograph of a thin section through small spicule on high-temperature spicular sinter that forms the inner part of the pool rim within the splash zone of a boiling spring that shows stromatolite-like growth laminations. Photo is approximately 0.15 cm across.

and Walter, 1976b, Figs. 22–24). Large accretionary particles, termed sinter pisoliths, form locally in splash zones (Walter, 1976b, Figs. 15–17).

3.1.4. Near-Vent Outflow Ponds and Basins

Between the main pools and outflow channels of many springs are broad, shallow depressions where water stands continuously (ponds) or for brief periods after eruptions or surges before flowing off the mound (basins). Examples include the shallow outflow pond at Steep Cone Spring (Fig. 4), which is located outside of the continuous circular rim of this boiling pool and is continuously full of water hotter than 73°C, and a similar pond at Octopus Spring (Fig. 7), which is separated from the main pool by a zone of discontinuous sinter islands, probably representing an older, flooded pool margin. Near-vent outflow basins are filled only during eruptions, when they may lie totally or partially within the splash and surge zone (Fig. 6).

Near-vent outflow ponds and basins are sites of extensive evaporation and of the deposition of a wide variety of sinter types in subaqueous to intermittently wet settings.

3.2. Proximal Outflow Channel

Where spring waters leave a pool or near-vent outflow pond, they are commonly collected within a single narrow channel or channels, if there is more than one outflow system on a mound (Fig. 10). At Clepsydra Geyser at Fountain Paint Pots, water splashed over a broad zone around the south side of the vent is collected into a single runout channel that rapidly narrows into a single channel less than 50 cm wide and more than 30 meters long, in which flow velocities exceed 1 m/s. Where this channel empties into another pool lower on the mound, water temperatures are still higher than 73°C. Similarly, Steep Cone Spring has a proximal outflow channel between the near-vent pond and low-temperature zone that is about 5 m long and 15–30 cm wide where flow velocities are generally higher than 50 cm/s and flow depths are less than 10 cm. There is relatively little silica deposition along proximal outflow channels, although evaporation at the air–water interface, often combined with splash and evaporation, has generally built small flanking sinter levees that keep the flows narrow and confined, reduce heat loss, and promote longer runout at high temperatures.

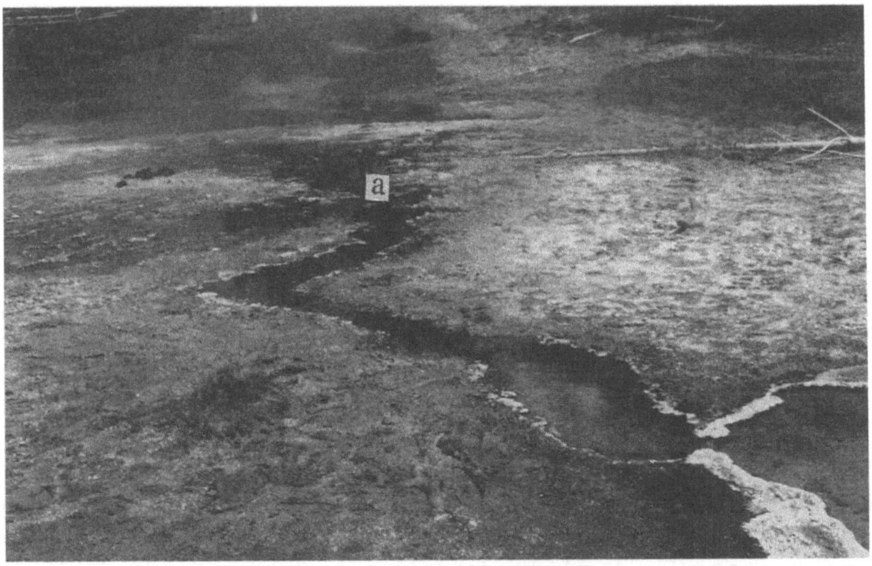

Figure 10. Proximal outflow channel at Octopus Spring. The spring, lower right, is characterized by temperatures hotter than 73°C and has a light bottom without obvious bacterial mats. The outflow channel has a medial proximal zone whose temperatures are above 73°C and a light gray bottom, but most has a temperature below 73°C, and the bottom is covered by dark *Synechococcus-Chloroflexus* mats. Note the paucity of the fresh sinter deposits along the proximal outflow channel. At "a," the proximal outflow channel bifurcates and spreads out to form the medial channel complex, most of which is covered by *Phormidium* mats.

Figure 11. Close-up of Steep Cone Spring showing the medial channel complex and associated *Phormidium* and *Calothrix* bacterial mats. The *Phormidium* mats show alternating medium gray streaks, representing actively growing orange to greenish *Phormidium* mats, and white areas, composed of white sinter that represents dead *Phormidium* mats from which organic matter has been removed. *Calothrix* mats occur lower on the slopes and flanking the *Phormidium* mats and are a uniform darker gray color in the photo and a grayish brown to tan color in the field. The *Calothrix* zone typically shows low terraces to fine terracettes (right).

3.3. Channel and Terrace Complex

In continuously active outflow systems, when the water temperature reaches 59–60°C, the wetted surface is covered by bacterial mats. These mats result in the buildup of medial "bars" and splitting of the outflow channel into a complex of distributaries. Where slopes are high, as at Steep Cone Spring, these distributaries are bordered by rubbery bacterial levees and overbank mats (Fig. 11). In low-gradient systems, channel splitting is accompanied by the formation of terraces and pools (Fig. 12). Sinter deposition in this zone is controlled by the pervasive bacterial mats and its structuring will be discussed in the following section on bacterial facies.

3.4. Sinter-Debris Apron

In the lower part of the medial channel complex, the outflow system tends to spread out, terraces disappear, and flow is concentrated into shallow channels separated by inactive parts of the mound covered by veneers of broken sinter and sinter debris (Figs. 3 and 11). Similar areas covered by sinter debris are developed higher on the mound between active parts of the outflow system. Flow usually occurs within very shallow channels that have hard sinter bottoms. In distal areas, the normal flow within individual channels is usually only a few millimeters to a centimeter or two deep, and the solid bed commonly

Figure 12. Terraces developed in the medial channel complex in the Clypsydra Geyser outflow system at Fountain Paint Pot. The high terraces are covered by *Phormidium*, and the lower terraces in the foreground are within the *Calothrix* zone.

shows transverse ripple-like ribs or microterracettes covered with brown *Calothrix* bacteria separated by flat, light gray areas of bare sinter (Fig. 11). These outflow channels are confined by debris accumulations and bifurcate and anastomose downflow. In many low-flow systems, the outflow eventually vanishes into the sinter debris (Fig. 3) whereas in others, discrete channels persist and empty into through-flowing streams or grade into grass-covered marshes and meadows.

3.5. Diatom Marsh and Meadow

Flat meadows surround many sinter mounds. Geothermal waters that flow off sinter mounds commonly enter these meadows to form marshes, where diatoms flourish (Fig. 3). Many marches surrounding sinter mounds are areas of silica deposition from the accumulation of diatomaceous debris. In addition, silica-charged waters commonly coat and silicify other plant structures, such as pine needles and plant roots. These can be buried and preserved in the geologic record. The areal extent of the diatom marsh and meadow deposits can exceed that of the mound itself.

4. TEMPERATURE AND BACTERIAL SUBDIVISIONS OF OUTFLOW SYSTEMS

The outflow system can also be subdivided into zones or facies based on the benthic bacteria living in the outflow waters. The types of bacteria living in the outflow system are, in turn, controlled largely by temperature. Four main bacterial/temperature zones can be recognized in the springs studied: (1) a high temperature bacterial biofilm zone (>73°C); (2) the *Synechococcus-Chloroflexus* zone (73–60°C); (3) the *Phormidium* zone (60–30°C); and (4) the *Calothrix* zone (<30°C). Although these zones can be characterized by the dominant type of bacteria, the actual benthic communities are complex associations of microorganisms that commonly include several types of prokaryotes, as well as algae and invertebrates. These associations include mixtures of organisms that inhabit the same mat layers, as well as stratified communities in which a succession of layers differing in available light, nutrients, oxygen levels, or other properties host different organism populations.

4.1. High Temperature Zone

Visible bacterial mats are absent in the high temperature (>73°C) zones of pools and outflow systems. The bottoms and sides of geysers and springs and high temperature parts of the outflow systems are composed of white, light gray, or grayish-brown sinter that lacks megascopically visible bacterial communities (Figs. 6 and 7). However, even extremely high temperature springs (>90°C) contain living bacteria (e.g., Brock, 1967; Brock et al., 1971), detected initially through experiments that sampled materials floating in the pools (Brock, 1967; Stahl et al., 1985). Recent studies confirmed the presence of thin bacterial biofilms on splash-zone sinter around high temperature springs (Cady, 1995), and it seems likely that the floating bacteria represent debris from benthic bacterial biofilms that coat the bottoms and sides of virtually all pools at even the highest temperatures.

Although silicified bacterial filaments have been reported in columnar and spicular splash-zone sinter within pools and near-vent areas at the ultramicroscopic level (Cady, 1995), it is as yet unclear whether the larger scale structuring of high temperature sinter reflects the presence of bacteria, as suggested by Cady (1995), or whether it is controlled by precipitative processes and hydrodynamic conditions, such as the degree of submergence, the intensity of splashing and surging, and the amount of evaporation. Walter (1976b) regarded high temperature sinter (geyserite) as strictly abiotic in origin. Our studies to date suggest that although bacteria may enhance silica deposition rates locally by providing increased surface area on a microscopic scale for silica precipitation, the larger scale (centimeter and above) structuring of sinter within the high temperature zone is related to spring and geyser hydrodynamics.

A variety of sparse high temperature benthic bacteria live within proximal outflow channels at temperatures higher than 73°C. Pink filamentous bacteria (Reysenbach et al., 1994) are present in proximal runout channels at temperatures as high as 83°C in the Octopus Spring area. We observed similar pale gray to yellowish filaments, possibly representing *Thermothrix*, in proximal runout channels where temperatures are as high as 80°C at Clepsydra Geyser. In general, these filamentous bacteria are not silicified, but at Steep Cone, bundles of apparently dead silicified pale yellowish filaments were seen in the proximal outflow channel at water temperatures higher than 73°C.

4.2. *Synechococcus-Chloroflexus* Zone

Between 73°C and 59–60°C, most pools and outflow systems are characterized by yellow, orange, and greenish bacterial biofilms and mats of *Synechococcus* and *Chloroflexus*. In general, surfaces in contact with outflow waters between 73°C and about 67–68°C are coated by thin biofilms of the yellow to greenish yellow unicellular coccoidal cyanophyte *Synechococcus sp.*, and surfaces in contact with waters between about 68°C and 60°C are covered by green to orange laminated to nodular mats that include the photosynthetic filamentous bacterium *Chloroflexus aurantiacus*, as well as *Synechococcus* (Castenholz and Pierson, 1995).

In hot springs with boiling (about 93–94°C at the altitude of Yellowstone) or near boiling waters, the upper part of the proximal outflow channel tends to carry waters hotter than 73°C. These channels show white to light tan sinter surfaces that lack obvious bacterial mats (Fig. 10), and generally have narrow marginal zones within which the temperature rapidly grades into surrounding ambient temperatures. The marginal zones show narrow linings of *Synechococcus* and *Chloroflexus* marking the appropriate temperature interval. The bottoms and sides of the lower parts of proximal outflow channels, where waters are <73°C and >60°C, are commonly completely covered with *Synechococcus* and *Chloroflexus* mats (Fig. 10 and Castenholz and Pierson, 1995, Fig. 2, p. 89, and color plate 1B).

Thin *Synechococcus* biofilms and *Chloroflexus* mats also coat the bottoms of many hot springs where the temperature is below 73°C (Walter, 1976c, Fig. 4, p. 493) and line the margins of high-temperature pools along the waterline where surface temperatures are 60–73°C.

It is not clear that the presence of *Synechococcus* biofilms or *Chloroflexus* mats substantially affects silica deposition. We have found no evidence that the mats and biofilms are themselves silicified, and there appear to be no distinctive siliceous deposits associated with *Synechococcus* biofilms and *Chloroflexus* mats (Figs. 7 and 10). Silica deposits within the *Synechococcus-Chloroflexus* zone are related largely to evaporation along pool and stream margins.

4.3. *Phormidium* Zone

Between 59–60°C and about 30°C, the outflow systems of continuously active alkaline springs host extensive bacterial mats dominated by species or strains of the filamentous cyanobacterial genus *Phormidium* (Copeland, 1936). This biological zone corresponds to the upper part of the medial channel complex. Mat growth in this part of the outflow system generally results in splitting the outflow channel into a complex of separate distributaries separated by quiet-water bacterial sheets and pools and, where the slope of the sinter surface is low, in the formation of terraces and pools (Fig. 12).

The entire active surface of the runoff system tends to be covered by mats, which exhibit a variety of morphologies in response to local hydrodynamic conditions. These include (1) active streams, which are characterized by longitudinal bacterial streamers, and associated flat rubbery sheets (Fig. 13), which are characterized by a complex bacterial community including *Mastigocladus laminosus*; (2) terrace pools, characterized by small subaqueous conical stromatolites and flat-topped stromatolites, both constructed by varieties of *Phormidium* (Fig. 14a); and (3) terrace fronts, characterized by thick, rubbery mats (Fig. 12). In addition to the dominant bacteria, *Chloroflexus aurantiacus* and *Synechococcus* are present in many of these settings (Walter et al., 1976).

All of these *Phormidium*-zone terrace and outflow channel complexes are thoroughly silicified and consist of tightly packed and intertwined bacterial filaments coated by thin films of silica. The active surfaces of the mats, which may be a few millimeters or less in thickness, are characterized by living bacteria without silica coatings. When any part of the

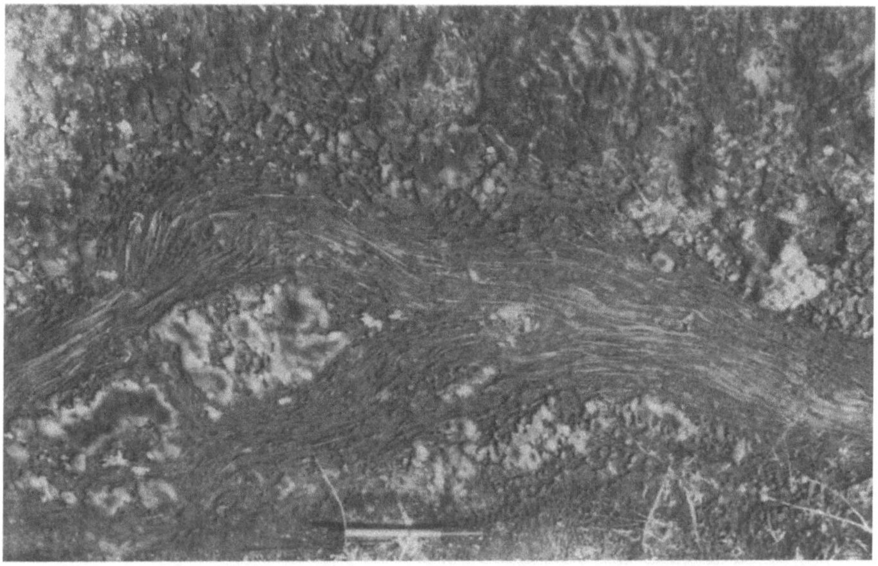

Figure 13. Plan view of the *Phormidium*-zone bacterial mats. The actively flowing stream is characterized by leathery streamers that include *Mastigocladus (Fischerella)*. Flanking quiet-water banks and shallow pools are characterized by conical and columnar varieties of *Phormidium*.

system is abandoned and dries, the bacteria die. In such exposed areas, the organic matter tends to degrade rapidly (Farmer et al., 1995), leaving a surface of white silica that retains the morphology and structure of the living mats (Fig. 14b) down to the level of hollow tubules that represent the silica coatings on individual filaments. The build-up of *Phormidium* mats occurs in large part through the vertical growth of bacterial filaments that form columnar and conical stromatolites. Much of the internal structuring of *Phormidium* sinter consists of flat, horizontal to gently dipping layers, from less than 1 cm to more than 10 cm thick that are composed of closely spaced vertical to subvertical silica tubules, generally about 5–25 microns in diameter, that represent silica coatings on individual bacterial filaments (Fig. 15). Layers of vertical silicified filaments are separated by thin, layers, up to a few millimeters thick, in which the filaments are oriented horizontally (Fig. 15).

Phormidium sinter is extremely low-density and very porous and commonly has primary porosities above 80%. Walking on *Phormidium* mats in pools or outflow channels can crush the delicate silica framework, causing one to sink up to 25–50 cm into the underlying sediment. In abandoned areas, *Phormidium* sinter tends to break up and degrade rapidly, due largely to trampling by bison, elk, and other mammals.

4.4. *Calothrix* Zone

Below about 30°C, *Phormidium* mats are replaced by grayish-brown flat to pustular mats dominated by filamentous cyanobacteria of the genus *Calothrix*. The *Calothrix* zone is dominated by low terraces or terracettes, and flow is spread evenly across the terraces with few main runoff channels (Figs. 11 and 12). Terrace pools are shallow, generally less than 5

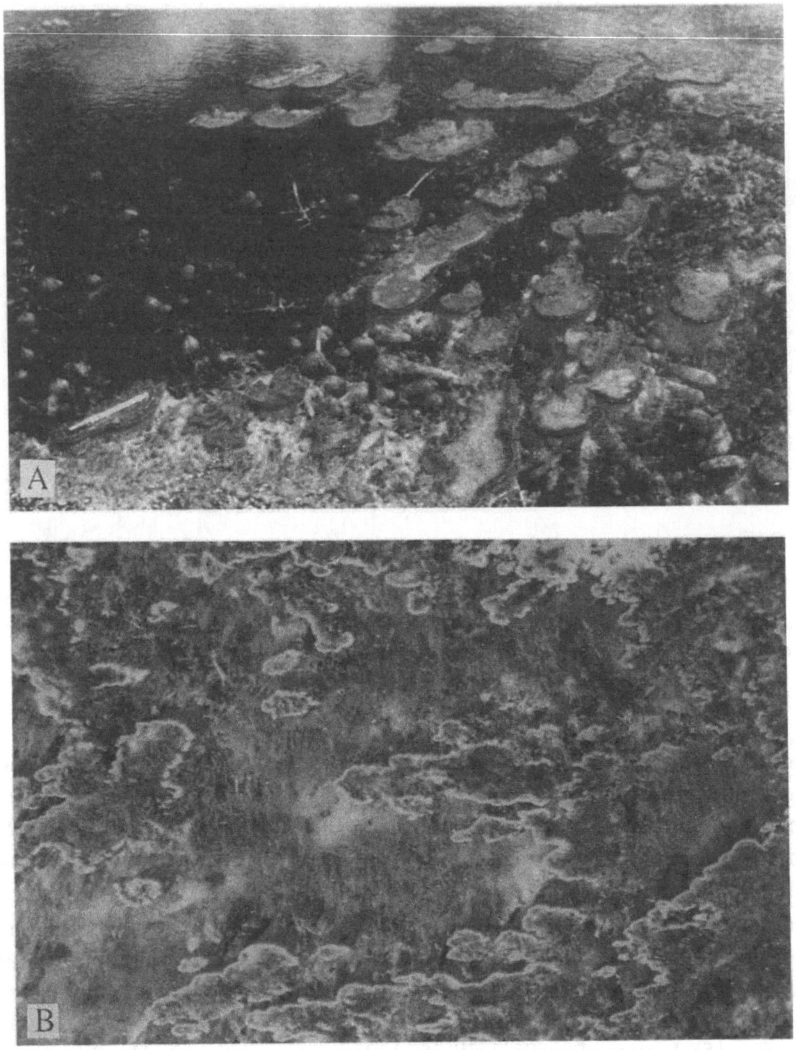

Figure 14. *Phormidium* zone mats and sinter. (a) Shallow pool containing small subaqueous conical and columnar flat-topped *Phormidium* stromatolites. Surfaces are covered by living mats, except in patches on some subaerial column tops (light gray). (b) Drained shallow pool as in (a). Mats have died and decayed, and the entire surface is composed of siliceous sinter.

cm deep. In some pools, *Calothrix* forms small microspicular branching shrub-like masses but most terrace fronts, tops, and pool bottoms are characterized by closely packed vertical to subvertical bacterial filaments. *Calothrix* sinter forms like *Phormidium* sinter; the individual bacterial filaments are coated by silica everywhere but at the very surface of the living mats. *Calothrix* sinter is denser than *Phormidium* sinter and is highly resistant to crushing. Sections through *Calothrix* sinter show stacked thin layers from 0.5 to 5 cm thick composed of vertical to subvertical silica-coated bacterial filaments. These layers are

Figure 15. Cross-sectional view of *Phormidium* ("p") and *Calothrix* ("c") sinters. *Phormidium* sinter is more porous and shows poorly developed conical stromatolites. *Calothrix* sinter consists mainly of thin layers of simple vertical to inclined silica filaments separated by thin continuous horizontal laminations.

separated by millimeter-thick layers of horizontal laminated silica or by thin zones in which the filaments are largely horizontal (Fig. 16). Because of its tightness and resistance to breakage, *Calothrix* sinter is commonly preserved and forms large parts of most sinter mounds.

In the zone of debris that characterizes the lower and inactive portions of sinter mounds, shallow, continuously wet runout channels show the spotty development of *Calothrix*, mainly as brownish tufts that line channel margins and filament bundles on transverse sinter ribs within the shallow channels. Sinter from this zone tends to break into thin plates, from 1 to 5 mm thick, showing layers of vertical to inclined silicified filaments between thin horizontal silica laminations. The flank deposits of Steep Cone Spring consist of layers of angular sinter detritus separated by beds of platy sinter showing layers 1–2 mm thick of silicified *Calothrix* filaments.

5. THE ROLE OF BACTERIA IN THE DEPOSITION AND STRUCTURING OF SILICEOUS SINTER

Previous studies of sinter formation have concluded that bacteria do not play an active role in silica deposition. Instead, bacterial filaments and coccoids act as surfaces upon which polymerized silica precipitates as a consequence of declining silica solubility (White et al., 1956). The results of this study largely reinforce this inference.

In the high temperature zone, >73°C, bacterial biofilms are widely present, and silica precipitation on bacterial filaments has been interpreted as exerting a strong influence on sinter microstructure, including the formation of stromatolite-like laminated columns and

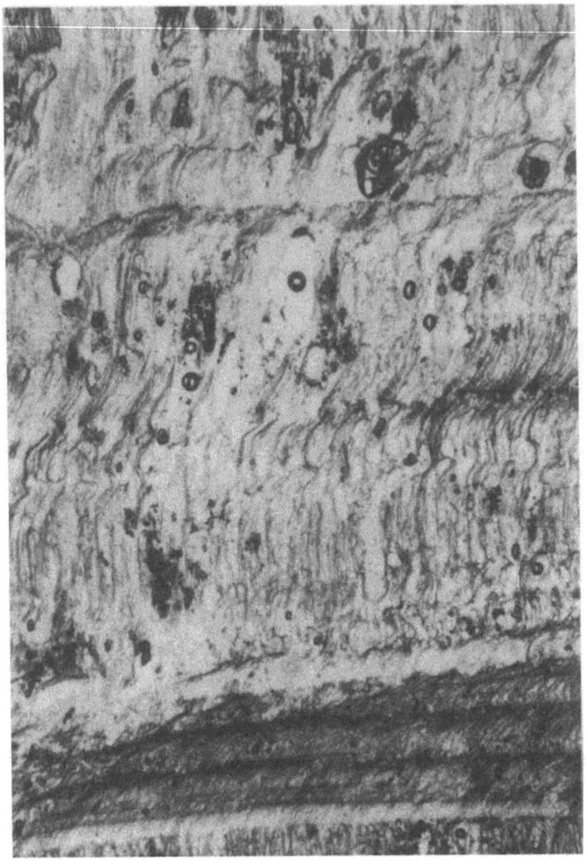

Figure 16. Photomicrograph of *Calothrix* sinter showing individual bacterial filaments. Filaments curve upward, tending to become more inclined at layer tops. The thin, horizontal laminations that separate layers of vertical to subvertical filaments (Fig. 15) are composed of tightly packed inclined filaments, filament ends, and some horizontal filaments.

spicules (Cady, 1995; Cady and Farmer, 1996; Jones et al., 1997). The rapid variation in sinter morphology and structuring at scales above 1 cm with pool/geyser hydrodynamics suggests to us that macrostructure of high-temperature sinter is controlled largely by abiological processes.

From 60–73°C, *Synechococcus* and *Chloroflexus* mats show little or no silicification. The mats or biofilms are thin, but even where *Chloroflexus* mats are well developed, we have not observed significant silicification. Although it might be suspected that minor siliceous deposits could form in association with these bacterial layers, they have not been observed.

Bacteria play a critical role in the deposition and structuring of sinter within the *Phormidium* and *Calothrix* zones below 60°C. The basic architectural elements of this system, including channels, terraces, and pools, probably reflect hydrodynamic conditions and can form through direct precipitation even in the absence of organisms, such as in rapidly aggrading carbonate systems and in cave deposits. However, the smaller macroscale features, including the layering and internal structure of *Phormidium* and *Calothrix* sinter, directly reflect the presence of bacteria (Walter et al., 1976; Cady and Farmer, 1996). The centimetric-scale layering, dominance of vertical filament fabrics, and a variety of minor structures all reflect the role of bacterial filaments and mats as templates for silica precipitation. There is no unambiguous evidence that the bacteria actively cause or mediate silica deposition, but it is possible that, by controlling the intramat chemical milieu, especially pH and the presence of organic acids, bacteria may play a more active role in silica precipitation than heretofore supposed, as suggested in the case of petrified wood (Leo and Barghoorn, 1976) and some early diagenetic marine cherts (Zijlstra, 1987).

To evaluate growth and silicification rates, clean plates of flat sinter were placed in several pools in the *Phormidium* facies of the Clepsydra Geyser runout system in May 1994. The plates were examined 1 month later and all had luxuriant growths of loosely packed, vertical to subvertical *Phormidium* filaments that form diffuse conical stromatolites up to 3 cm high, none of which showed any silicification. Mat growth is clearly independent of silicification and can outstrip silicification during rapid colonization of new surfaces. We suspect that as the bacterial growth rate declines with crowding, silicification cements the filaments, except for the actively growing filament tips.

Silicification in the diatom marsh and meadows commonly involves the precipitation of opaline coatings on *in situ* plant stems and roots and on fallen leaves and needles. These biological materials were also passive templates for silica precipitation, although the presence of organic acids in soils may promote silica precipitation. Diatoms actively precipitate silica to form frustules, and these comprise fine sediments, commonly mixed with mud and organic matter, in marshes surrounding hot springs.

6. DISCUSSION

This study provides a framework for subdividing sinter mounds and their outflow systems in Yellowstone into major morphological and biological zones from geysers and boiling pools at temperatures of 93–94°C to surrounding diatom marshes and meadows at ambient temperatures. Each morphological zone is characterized by a temperature spectrum and an associated benthic bacterial population. Silica deposition occurs in all zones, but the morphology, structuring, and role of bacteria in sinter deposition is different in each. In highly active springs, water temperatures are commonly near boiling, and the outflow systems show sinter, temperature, and bacterial zones that reflect the progressive decrease in water temperature from 90–94°C to ambient surface temperatures. In the high temperature zone, >73°C, benthic bacterial biofilms are widely and probably ubiquitously present and have been interpreted as influences on microscopic structuring of the sinter deposits. Larger scale (>1 cm) sinter structuring is controlled largely by direct precipitation and local hydrodynamic conditions, especially the amount of splash and evaporation.

Between 73°C and 60°C, bacterial biofilms and mats composed of *Synechococcus* and *Chloroflexus* are present but are associated with little sinter deposition and play no role in sinter deposition and structuring. Below 60°C, bacterially-influenced sinter predominates, and mat growth promotes splitting of outflow channels and formation of the channel/distributary system characteristic of the medial channel complex. The large-scale structures, such as terraces, channels, and pools, are probably hydrodynamically controlled, but finer scale structuring reflects the presence of bacterial mats. The predominant structures are layers of vertical filaments and filament clusters that form conical stromatolites. These are preserved through the precipitation of amorphous opaline coatings, and the bacteria apparently are mainly passive surfaces upon which precipitation occurs. However, the possibility that silica precipitation is influenced or controlled within the mat by mat-regulated water chemistry, as in the case of some cherts and petrified wood, also needs to be investigated. Thin dense laminations of inclined to horizontal filaments separate the thicker, more porous layers of vertical filaments. Small, sharply-pointed conical stromatolites are commonly developed in the *Phormidium* facies (Walter et al., 1976).

Many springs have vent temperatures below those described above. The bacterial populations in these systems reflect existing temperature gradients, not pool morphological zones. Where vent temperatures are near or below 73°C, virtually all morphological zones of the pools, which may have been established under higher temperature conditions, will be covered with *Synechococcus-Chloroflexus* or even *Phormidium* mats and the outflow system will show correspondingly low-temperature mats. Hence, as pools evolve and their activity declines, there is a tendency for lower temperature bacterial/temperature zones to migrate up-system and replace the high temperature zones that may have characterized the springs during their most active stages.

This general morphological, depositional, and biological zonation is applicable to most sinter mounds within Yellowstone in continuously flowing hot springs and geysers. In some areas, however, the temperature structure and/or morphological features result in the poor development of some bacterial zones and the poor development or absence of some morphological zones. For instance, proximal outflow channels are absent in some springs and the vent pool grades directly into the medial channel complex, commonly accompanied by a compression of the *Synechococcus-Chloroflexus* zone. Episodic springs and geysers show different divisions. Most lack the morphological zones outlined here and, in systems that dry between eruptions, obvious bacterial mats are absent (Fig. 6). Zonation and facies of these systems will be considered in a separate report.

Postdepositional alteration has a major effect on sinter structuring. Burial of as little as 20 centimeters below the surface of an active sinter mound exposes the porous opaline sinter to hot hydrothermal fluids moving upward from subsurface sources and outward through the porous sinter from active vent feeder channels. Cooling of the subsurface fluids results in silica precipitation within the open pore spaces, commonly forming layers of nearly solid opaline silica from layers of vertical bacterial filaments. The filamentous structure is largely obliterated during such early, shallow diagenesis. Preservation of bacterial structures in the geologic record, in the absence of preserved carbonaceous remains, probably requires an interval during which the initially open pore spaces can be filled with some other material or the silicified filaments can be coated with iron oxide or similar impurities. Later silica cementation or recrystallization produces rocks with textural and/or compositional contrasts between the filament coatings and cavity fill and allows distinguishing the filaments and fill visually and compositionally.

7. SUMMARY

The study of siliceous sinter deposited around modern hot springs provides an important means of evaluating the morphology, structuring, and genesis of siliceous deposits formed under a wide variety of temperature, hydrodynamic, and biological conditions. It is hoped that an understanding of the complex relationships among modern hot springs hydrodynamics, sinter morphology and structuring, and thermophilic bacteria will lead to better criteria for identifying ancient hot spring deposits in the geologic record, for determining the biogenicity of possible bacterial structures in early Precambrian rocks, and for interpreting the nature, distribution, and paleoecology of ancient organisms.

ACKNOWLEDGMENTS. This research has been supported by grant NCC 2-721 from the NASA Exobiology Program. We are also grateful to the Research Office at Yellowstone National Park, especially Mr. John Varley, Mr. Robert Lindstrom, and Ranger Les Inafuku for permission to work in Yellowstone and for invaluable help and advice in the field.

REFERENCES

Brock, T. D. 1967. Life at high temperatures. *Science* **158:**1012.

Brock, T. D., and Starr, M. P. 1978. *Thermophilic microorganisms and life at high temperatures.* New York: Springer-Verlag.

Brock, T. D., Brock, M. L., Bott, T. L., and Edwards, M. R. 1971. Microbial life at 90°C: The sulfur bacteria of Boulder Spring. *J. Bacteriol.* **107:**303.

Byerly, G. R., Lowe, D. R., and Walsh, M. M. 1986. Stromatolites from the 3,300–3,500-Myr Swaziland Supergroup, Barberton Mountain Land, South Africa. *Nature* **319:**489.

Cady, S. L. 1995. Columnar and spicular geyserites from Yellowstone National Park, WY; Scanning and transmission electron microscopy evidence for biogenicity. *Geol. Soc. Am. Abstracts with Programs* **27:**A305.

Cady, S. L., and Farmer, J. D. 1996. Fossilization processes in siliceous thermal springs: Trends in preservation along thermal gradients. In Bock, G. R., and Goode, J. A. (eds.), *Evolution of hydrothermal ecosystems on Earth (and Mars?)* (pp. 150–173). Chichester: Wiley.

Castenholz, R. W., and Pierson, B. K. 1995. Ecology of thermophilic anoxygenic phototrophs. In Blankenship, R. E., Madigan, M. T., and Bauer, C. E. (eds.), *Anoxygenic photosynthetic bacteria* (pp. 87–103). Dordrecht, The Netherlands: Kluwer Academic.

Christiansen, R. L. 1984. Yellowstone magmatic evolution: Its bearing on understanding large-volume explosive volcanism. In *Explosive volcanism, inception, evolution, and hazards* (pp. 84–95). Washington, DC: National Academy Press.

Copeland, J. J. 1936. Yellowstone thermal Myxophyceae. *Ann. N.Y. Acad. Sci.* **36:**1.

Dill, R. F., Kendall, C. G. St. C., and Shinn, E. A. 1986. Giant subtidal stromatolites forming in normal salinity water. *Nature* **324:**55.

Farmer, J. D., Cady, S. L., and DesMarais, D. J. 1995. Fossilization processes in thermal springs. *Geol. Soc. Am. Abstr. with Programs* **27:**A305.

Hoffman, P. 1976. Stromatolite morphogenesis in Shark Bay, Western Australia. In Walter, M. R. (ed.), *Stromatolites* (pp. 261–272). Amsterdam: Elsevier.

Jones, B., Renaut, R. W., and Rosen, M. R. 1997. Biogenicity of silica precipitation around geysers and hot-spring vents, North Island, New Zealand. *J. Sedim. Petrol.* **67:**88.

Leo, R. F., and Barghoorn, E. S. 1976. Silicification of wood, Harvard University. *Botanical Museum Leaflets* **25**(1)**:**1.

Lowe, D. R. 1980. Stromatolites 3,400-Myr old from the Archaean of Western Australia. *Nature* **284:**441.

Lowe, D. R. 1994. The abiological origin of described stromatolites older than 3.2 Ga. *Geology* **22:**387.

Maliva, R. G., Knoll, A. H., and Siever, R. 1989. Secular changes in chert distribution: A reflection of evolving biological participation in the silica cycle. *Palaios* **4:**519.

Morgan, W. J. 1972. Convection plumes and plate motions. *Am. Assoc. Pet. Geol. Bull.* **56:**203.

Pierce, K. L., and Morgan, L. A. 1992. The track of the Yellowstone hot spot: Volcanism, faulting, and uplift. In Link, R. K., Kuntz, M. A., and Platt, L. B. (eds.), *Regional geology of eastern Idaho and western Wyoming* (pp. 1–53). *Geol. Soc. Am. Memoir* 179.

Playford, P. E., and Cockbain, A. E. 1976. Modern algal stromatolites at Hamelin Pool, a hypersaline barred basin in Shark Bay, Western Australia. In Walter, M. R. (ed.), *Stromatolites* (pp. 389–412). Amsterdam: Elsevier.

Reysenbach, A.-L., Wickham, G. S., and Pace, N. R. 1994. Phylogenetic analysis of the hyperthermophilic pink filament community in Octopus Spring, Yellowstone National Park. *Appl. Environ. Microbiol.* **60:**2113.

Schopf, J. W. 1975. Precambrian paleobiology: Problems and perspectives. *Annu. Rev. Earth Planet. Sci.* **3:**213.

Schopf, J. W. 1976. Are the oldest "fossils" fossils? *Origins of Life* **7:**19.

Schopf, J. W. 1983. Archaean microfossils: New evidence of ancient microbes. In Schopf, J. W. (ed.), *Earth's earliest biosphere* (pp. 214–239). Princeton, NJ: Princeton University Press.

Siever, R. 1992. The silica cycle in the Precambrian. *Geochim. Cosmochim. Acta* **56:**3265.

Stahl, D. A., Lane, D. J., Olsen, G. J., and Pace, N. R. 1985. Characterization of a Yellowstone hot spring microbial community by 5S rRNA sequences. *Appl. Environ. Microbiol.* **49:**1379.

Suppe, J., Powell, C., and Berry, R. 1975. Regional topography, seismicity, quaternary volcanism, and the present-day tectonics of the western United States. *Am. J. Sci.* **275A:**397.

Thrailkill, J. 1976. Speleothems. In Walter, M. R. (ed.), *Stromatolites* (pp. 73–86). Amsterdam: Elsevier.

Walter, M. R. 1976a. Introduction. In Walter, M. R. (ed.), *Stromatolites* (pp. 1–3). Amsterdam: Elsevier.

Walter, M. R. 1976b. Geyserites of Yellowstone National Park: An example of abiogenic "stromatolites." In Walter, M. R. (ed.), *Stromatolites* (pp. 87–112). Amsterdam: Elsevier.

Walter, M. R. 1976c. Hot-spring sediments in Yellowstone National Park. In Walter, M. R. (ed.), *Stromatolites* (pp. 489–498). Amsterdam: Elsevier.

Walter, M. R., Bauld, J., and Brock, T. 1976. Microbiology and morphogenesis of columnar stromatolites (*Conophyton, Vacerrilla*) from hot springs in Yellowstone National Park. In Walter, M. R. (ed.), *Stromatolites* (pp. 273–310). Amsterdam: Elsevier.

Walter, M. R., Buick, R., and Dunlop, J. S. R. 1980. Stromatolites 3,400–3,500 Myr old from the North Pole area, Western Australia. *Nature* **284:**443.

White, D. E., Brannock, W. W., and Murata, K. J. 1956. Silica in hot spring waters. *Geochim. Cosmochim. Acta* **10:**27.

White, D. E., Thompson, G. A., and Sandberg, C. H. 1964. Rocks, structure, and geologic history of Steamboat Springs thermal area, Washoe County, Nevada. USGS Prof. Paper 458-B. 63 p.

Zijlstra, H. J. P. 1987. Early diagenetic silica precipitation, in relation to redox boundaries and bacterial metabolism, in late Cretaceous chalk of the Maastrichtian type locality. *Geol. Mijnbouw* **66:**343.

12

Use of 16S rRNA, Lipid, and Naturally Preserved Components of Hot Spring Mats and Microorganisms to Help Interpret the Record of Microbial Evolution

David M. Ward, Mary M. Bateson, and Jan W. de Leeuw

---

1. INTRODUCTION

Our knowledge of the history of microbial life on earth is based on several types of information, most of which is found in the geological record. The occurrence of stromatolites suggests that photosynthetic microbial mat communities, their most likely modern analogs, are as old as 3.5 billion years ago and were likely to have been the earth's predominant biota during the Precambrian period (Schopf, 1992a; Walter et al., 1992). Many stromatolites contain microfossils, and many of these resemble cyanobacteria and green nonsulfur bacteria which are known to form mats on modern earth (Schopf, 1992b). It has even been considered that microfossil evidence may support the presence of cyanobacteria (and thus oxygen-evolving photosynthesis) in the oldest stromatolites (Schopf, 1992a). Precambrian sedimentary rocks also contain chemical fossils, remnants of the original structural components of the microbial cells which biosynthesized them (Summons and Hayes, 1992). Most sedimentary organic matter is in a complex insoluble form (Strauss et al., 1992a), but some is present in carbon skeletons that can be related to modern cell constituents (Summons, 1992). Such organic remains may even contain a stable isotope

---

David M. Ward and Mary M. Bateson • Department of Microbiology, Montana State University, Bozeman, Montana 59717. **Jan W. de Leeuw** • Netherlands Institute for Sea Research (NIOZ), Division of Marine Biogeochemistry, 1790 AB Den Burg, Texel, The Netherlands.

Thermophiles: Biodiversity, Ecology, and Evolution, edited by Reysenbach *et al.* Kluwer Academic / Plenum Publishers, New York, 2001.

signature that informs us about the physiology of microorganisms in the Precambrian (Strauss et al., 1992b). For example, stable carbon isotopic signatures of organic carbon suggest that autotrophic biochemistry was important as early as 3.8 billion years ago (Mojzsis et al., 1996). Some microbial physiologies also left a geochemical signature in Precambrian rocks. The best example is the banded iron formations that, it is thought, represent massive iron oxidation following the evolution of physiologies that might have accomplished this (e.g., via oxygen originating from oxygenic photosynthesis or directly by bacterial photosynthesis using iron as a reductant) (Klein and Beukes, 1992; Widdel et al., 1993). Most recently, molecular phylogeny, largely based on comparative analysis of small subunit ribosomal RNA (16S and 18S rRNA) sequences of diverse extant organisms, provides a means of inferring evolutionary histories of microorganisms (Olsen et al., 1994; Woese, 1987) (Fig. 1). From this approach, we have learned of three major domains of evolutionary descent, two of which contain the known prokaryotes and one of which contains all enucleated cellular life forms (protozoa, algae, fungi, plants, and animals).

For this volume, it is appropriate to ask why the study of Yellowstone microbial communities is important to microbial evolution? A lesson from molecular phylogenetic studies which is emphasized in Fig. 1a is that thermophily is likely to be an ancestral phenotype (Woese, 1987; Pace, 1991). This is consistent with the suggestion that the

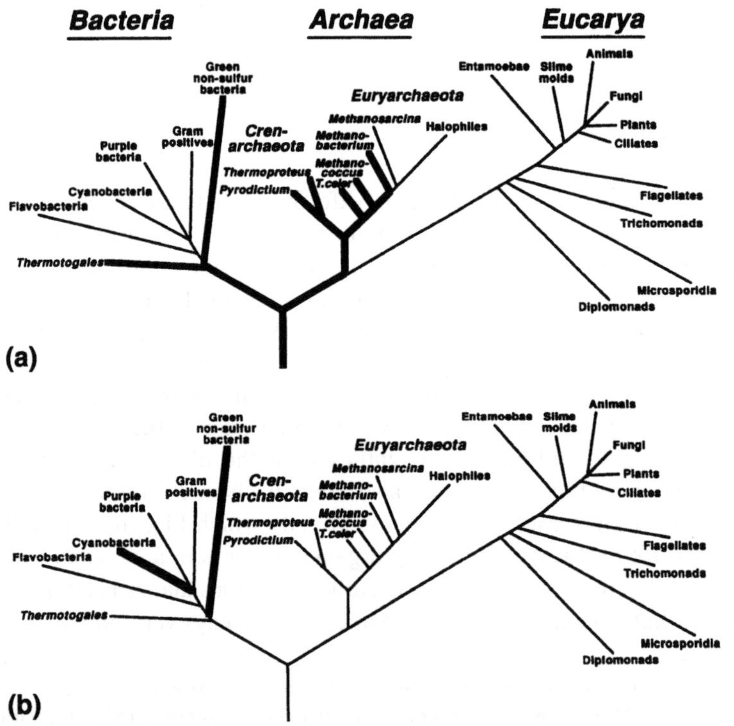

Figure 1. Distance matrix phylogenetic trees of representative major lines of descent highlighting (a) the thermophilic character of ancient bacteria and archaea (b) the relative positions of major photosynthetic lineages. Modified from Woese (1994).

environment of the early earth may have been extremely hot (Kasting and Chang, 1992). In fact, many bacterial and archaeal species that descended from such ancient ancestors have been detected in geothermal features of Yellowstone (Barns et al., 1994, 1996; Reysenbach et al., 1994; Pace, 1997). Continued investigation of natural thermal environments, especially using molecular methods, will provide further insights into the biodiversity of extant species, its ecological basis, and its evolutionary history. A second lesson from molecular phylogenetic studies is that photosynthetic microorganisms known to form mats on modern earth have different evolutionary histories. As depicted in Fig. 1b, green nonsulfur bacteria, like *Chloroflexus*, evolved from a common ancestor that is likely to have predated the common ancestor of cyanobacteria. Algae arose yet later and in the eucaryotic lineage as a product of horizontal gene flow because chloroplasts were the products of endosymbiosis between eucaryotic cells and cyanobacteria (Giovannoni et al., 1988; Bhattacharya and Medlin, 1995). Most modern photosynthetic mats occur in extreme environments, such as hypersaline pools and hot springs, where modern herbivorous animals are restricted (Pierson, 1992). Most mats occur in hypersaline ponds near coastlines, seemingly relevant to the coastal settings of many stromatolites (Schopf, 1992b). Hypersaline mats are constructed by a mixture of all three evolutionary groups of phototrophic microorganisms (D'Amelio et al., 1989; Des Marais et al., 1992b). However, mats in hot springs of differing chemistry are constructed by discrete groups of phototrophic microorganisms (Ward et al., 1989b). For instance, some mats in sulfide springs of near neutral pH, like "New Pit" Spring of Mammoth Terraces, Yellowstone National Park, are constructed exclusively by green nonsulfur bacteria. Mats in nonsulfidic alkaline springs like Octopus Spring of the Lower Geyser Basin, Yellowstone National Park, are constructed by cyanobacteria in conjunction with anoxygenic photosynthetic bacteria. Mats in acidic springs, like Nymph Creek, a few miles north of Norris Geyser Basin, Yellowstone National Park, are constructed by algae in the absence of cyanobacteria and anoxygenic photosynthetic bacteria. This makes it possible to evaluate whether different photosynthetic groups impart unique chemical signatures that might be recognized in the biochemical components of cells. Such signatures might even persist in the fossil record, which might record a progression in the evolution of photosynthetic mat types, as predicted by the molecular phylogenetic record.

The purpose of this chapter is to illustrate how we have examined different types of cell components in different types of Yellowstone microbial mat communities. Since 1977, we have studied these mats as models in which to investigate principles of microbial community ecology (see Ward et al., 1984, 1987, 1989a,b, 1992, 1994a, 1998; Ward and Castenholz, in press; Ferris et al., this volume), building upon a solid base of knowledge laid down by Brock, Castenholz and others (see Brock, 1978). Here, we provide reference to our published work on rRNA and lipid biomarkers and highlight some examples of our more recent findings of relevance to the evolution of microorganisms that inhabit hot spring microbial mats.

2. 16S rRNA BIOMARKER STUDIES LINK BIODIVERSITY, ECOLOGY, AND EVOLUTION

In the preface of their ecology text, Begon et al. (1990) made an important point by extending the famous remark of T.H. Dobzhansky that "nothing in biology makes sense, except in the light of evolution" with the comment that "very little in evolution makes sense, except in the light of ecology." This holds, of course, because evolution has occurred in response to natural selection—how the environment has shaped the origin of species. In recent years, the field of microbial ecology has been most fortunate. New molecular methods have been developed for the use of a cell component, 16S rRNA, whose sequence of over 1500 nucleotides constitutes a highly informative signature molecule for native populations of bacteria (Pace et al., 1986; Ward et al., 1992). Because this cell component has also enabled a rational view of microbial evolution (Fig. 1), it provides a means of detecting native microorganisms and also of understanding their evolutionary history. Ferris et al. (this volume) detail our current understanding of the biodiversity of native populations in hot spring mats. The view provided by rRNA approaches is indeed very different from the view provided by culture methods because almost none of the 16S rRNA sequences detected represent species cultivated from the mats. In an early critique of rRNA methods, Brock (1987) pointed out that "these studies have been done in the almost complete absence of ecological interpretation" and, undoubtedly because of the reasonable link between ecology and evolution cited before, "how much more satisfying would be our understanding of microbial evolution if we could relate it to microbial ecology." Here, we focus on what we have learned about the connections between the ecology and evolution of two major groups of mat phototrophs, cyanobacteria and green nonsulfur bacteria, whose 16S rRNA signatures we have readily detected.

2.1. Cyanobacterial Diversity, Ecology, and Evolution

The Octopus Spring cyanobacterial mat, it was formerly thought, is constructed by a cosmopolitan unicellular cyanobacterium, *Synechococcus lividus*. However, we now know of at least eleven different cyanobacterial 16S rRNA signatures in this mat (Ferris et al., 1996a,b, 1997, this volume; Ferris and Ward, 1997; Ward et al., 1990; Weller et al., 1992). It is rational to infer that these populations all conduct oxygenic photosynthesis because all oxygenic phototrophs constitute a monophyletic lineage in the domain Bacteria (Fig. 1). Two patterns can be observed in the evolution of these organisms. As shown in Fig. 2a, the genetic diversity among the Octopus Spring cyanobacteria spans the diversity among cyanobacteria from all habitats whose 16S rRNA sequences have been compared (Ferris et

Figure 2. Distance matrix phylogenetic trees illustrating the representative major lines of descent of (a) cyanobacteria, and (b) green nonsulfur bacteria and relatives. Thick lines are populations detected in the Octopus Spring cyanobacterial mat or the "New Pit" Spring *Chloroflexus* mat. The trees were constructed using the programs DNADIST, FITCH, SEQBOOT, and CONSENSE from the Phylogenetic Inference Package (PHYLIP) version 3.57c. Trees were inferred from nucleotides that align with *Escherichia coli* positions 332 to 452, 480 to 452, 480 to 507, 712 to 892, and 1140 to 1364 (a) and 256 to 446, 485 to 598, 604 to 830, 856 to 922, 933 to 939, and 954 to 966 (b), except that for sequences shown to the right of arrows, fewer nucleotides were available for analysis. For these cases, smaller trees were inferred from available data, and manually added at the appropriate branch. The cyanobacterial tree (a) was rooted by using the 16S rRNA sequences of *Thermotoga maritima*, *Chlorobium vibrioforme*, and *E. coli*, whereas the tree in (b) was rooted using sequences from *Methanobacterium formicicum*, *Thermodesulfobacterium commune*, *Aquifex pyrophilus*, *T. maritima*, *C. vibrioforme*, *Agrobacterium tumefaciens*, *Pseudomonas testosteroni*, and *E. coli*. The consensus values at the forks indicate the number of times a group consisting of the species to the right of the fork occurred among 100 trees inferred from the bootstrapped data sets. Evolutionary distance is indicated by horizontal lines; each bar corresponds to 0.01 fixed point mutations per sequence position.

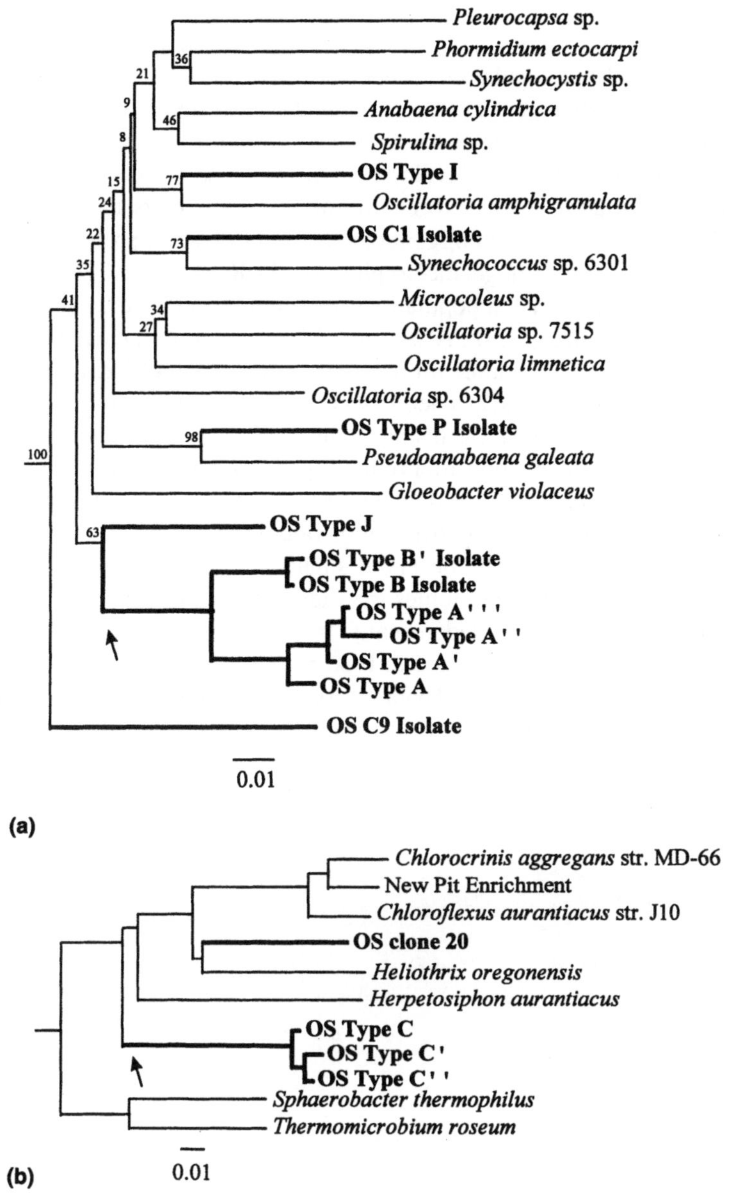
Pleurocapsa sp.
Phormidium ectocarpi
Synechocystis sp.
Anabaena cylindrica
Spirulina sp.
OS Type I
Oscillatoria amphigranulata
OS C1 Isolate
Synechococcus sp. 6301
Microcoleus sp.
Oscillatoria sp. 7515
Oscillatoria limnetica
Oscillatoria sp. 6304
OS Type P Isolate
Pseudoanabaena galeata
Gloeobacter violaceus
OS Type J
OS Type B' Isolate
OS Type B Isolate
OS Type A'''
OS Type A''
OS Type A'
OS Type A
OS C9 Isolate
0.01
(a)
Chlorocrinis aggregans str. MD-66
New Pit Enrichment
Chloroflexus aurantiacus str. J10
OS clone 20
Heliothrix oregonensis
Herpetosiphon aurantiacus
OS Type C
OS Type C'
OS Type C''
Sphaerobacter thermophilus
Thermomicrobium roseum
0.01
(b)

Figure 2.

al., 1996b). Though the rapid divergences among cyanobacteria make it difficult to document branching orders through bootstrap analysis, some of the Octopus Spring cyanobacteria may be the most deeply divergent cyanobacteria so far discovered. This would be consistent with the possibility that ancestral cyanobacteria may have inhabited thermal habitats.

A second pattern of evolution among Octopus Spring cyanobacteria is the occurrence of a cluster of highly related populations (populations A, A′, A″, A‴, B, and B′ in Fig. 2a). To link the evolutionary divergence of these populations with possible ecological driving forces, we need to know something about the phenotypes of the populations. Fortunately, the sequence uniqueness enables the development of strategies to probe the distribution and behavior of each population relative to environmental parameters and changes. For instance, there is good evidence from oligonucleotide hybridization probe (Ruff-Roberts et al., 1994) and denaturing gradient gel electrophoresis (DGGE) studies (Ferris et al., 1996a) to suggest that these highly related populations are uniquely adapted to different environmental situations. Populations B, B′, A, A′, and A″ are found at increasing temperatures from ca. 50°C to the upper temperature limit of the mat near 72–74°C, respectively (Fig. 3). At ca. 60°C, population A is found at a depth of ca. 400 μm beneath the mat surface, where population B′ is most abundant (Ramsing et al., 2000). Furthermore, population A‴ appears, as if a colonist species, soon after removing the cyanobacterial mat layer (Ferris et al., 1997). The combined ecological and evolutionary patterns imply that specialization to environmental parameters (such as narrow temperature ranges) has resulted in very little change in the 16S rRNA gene. The observed small-scale evolutionary radiation among thermophilic cyanobacteria thus resembles the radiation among Darwin's Galapagos finches, closely related as they evolved recently from the same common ancestors. It is hypothesized that the finches specialized to food resources; here, cyanobacterial populations have specialized to physiochemical parameters (and perhaps as colonists). Temperature-adapted *S. lividus* populations were previously cultivated from a hot spring cyanobacterial

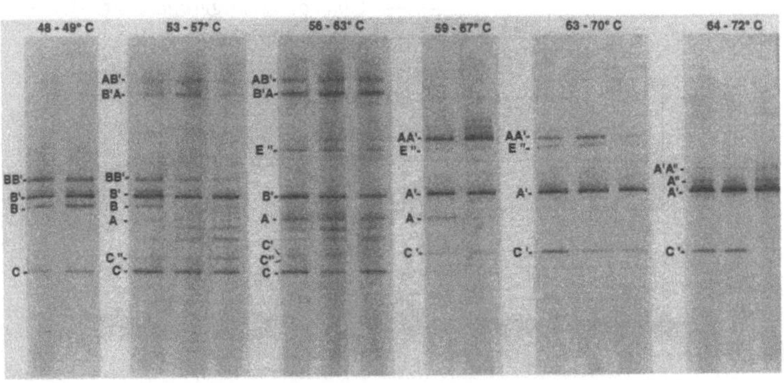

Figure 3. Denaturing gradient gel electrophoresis profiles of PCR-amplified 16S rRNA gene segments, indicating the temperature distribution of cyanobacterial (A, A′, A′, B and B′), green nonsulfur bacteria-like (C, C′ and C″), and green sulfur bacteria-like (E″) populations in Octopus Spring mat samples collected 13 March 1995 (from Ferris and Ward, 1997). Bands were identified by sequence analysis; bands with two-letter labels are heteroduplex artifacts.

mat (Peary and Castenholz, 1964), but the genetic relatedness of these populations was unknown. Considering the strong propensity for *S. lividus* to be selected in enrichment culture (Ferris et al., 1996b), we speculate that these, like the B, B′, A, A′, and A″ populations, may have been closely related. Adaptation to light is also likely because we have observed differential responses to light in oligonucleotide hybridization probe (Ruff-Roberts et al., 1994) and motility (Ramsing et al., 1997) studies. The forces that drove deeper evolutionary divergences among thermophilic cyanobacteria (e.g., among the A/B-like cluster and types I, J. P, C9 and C1) remain unknown. Adopting the view of evolutionary ecology for macroorganisms, one thinks of either other environmental adaptations or of geographic isolation as two major possibilities.

We have previously suggested that multiple specialized populations of functionally similar species fit the macroecological concept of guilds and that intraguild diversity might help stabilize guild activity in a changing environment (Ward et al., 1994). This would constitute an important emergent property of community structure, which arose as a consequence of the evolution of these diverse populations.

2.2. Diversity, Ecology, and Evolution of Green Nonsulfur-like Bacteria

The Octopus Spring mat is known to contain the green nonsulfur bacterium *Chloroflexus aurantiacus*. However, we now know of different green nonsulfur bacteria-like 16S rRNA signatures (Ferris et al., 1996a, this volume; Ferris and Ward, 1997; Ruff-Roberts et al., 1994; Ward et al., 1990; Weller et al., 1992). As shown in Fig. 2b, trends observed for cyanobacterial diversity were also observed for the green nonsulfur bacteria. The five signatures (including *C. aurantiacus*, recognized by cultivation [Pierson and Castenholz, 1974] and oligonucleotide hybridization probing [Ruff-Roberts et al., 1994]) span the diversity found among known members of this phylogenetic group, and a cluster of highly related populations (types C, C′, C″) exists. The closely related populations, which are those most readily detected by molecular means, exhibit temperature distributions which suggest specialization to distinct temperature ranges (Fig. 3). In this case, it is not possible to ascertain whether any of the four populations detected by molecular analysis are, like *C. aurantiacus*, photosynthetic bacteria, because the phylogenetic lineage also contains aerobic chemoorganotrophic species (e.g., *Herpetosiphon*, *Sphaerobacter*, *Thermomicrobium*). We need to determine the phenotypic character of these *Chloroflexus* phylogenetic relatives to unravel the evolutionary history and ecology of this group. If all of these populations are green nonsulfur bacteria, the similarity of the evolutionary patterns to those of cyanobacteria raises the intriguing possibility of coevolution between specific populations of these two physiological groups. This is logical because cyanobacteria are known to cross-feed products of photosynthesis to filaments that resemble *Chloroflexus*. For instance, glycolate produced during photorespiration (Bateson and Ward, 1988) and fermentation products which may result from dark fermentation of stored polyglucose (Nold and Ward, 1996) are readily photoassimilated by filamentous mat inhabitants (Bateson and Ward, 1988; Anderson et al., 1987). Such cross-feeding may involve transfer of the majority of carbon fixed in cyanobacterial photosynthesis (Nold and Ward, 1996). It is possible, for example, that populations of cyanobacteria and green nonsulfur bacteria, which are specialized to similar environmental parameters, coevolved. An alternative possibility is that these two groups underwent convergent evolution.

2.3. *Chloroflexus* sp. of Sulfidic Mats Is Closely Related to *C. aurantiacus*

In the sulfidic "New Pit" Spring a unique *Chloroflexus* sp., which is obligately phototrophic, constructs the mat, it is thought, in the absence of cyanobacteria (Giovannoni et al., 1987). It was of interest to determine the evolutionary history of the organism that constructed this mat because it might represent a more primordial species (i.e., one that diverged deeper in the tree than species like *C. aurantiacus*, which are associated with the more recently evolved cyanobacteria). We enriched a predominant *Chloroflexus* sp. from this spring by using the extincting dilution enrichment method (Ferris et al., 1996b). Its 16S rRNA signature was identical to the predominant 16S rRNA signature of the mat itself, as determined by DGGE (Ward et al., 1997). As shown in Fig. 2b, this organism is closely related (95% similar 16S rRNA sequences) to *C. aurantiacus* and does not branch more deeply than *C. aurantiacus*, indicating that it is not a more primordial green nonsulfur bacterium.

3. LIPID BIOMARKER STUDIES HELP LINK CHEMICAL FOSSILS TO THEIR MICROBIAL SOURCES

Classically, lipid cell components have been studied as biomarkers because lipid-like molecules occur in the chemical fossil record (Summons and Hayes, 1992). The difficulty has been in interpreting the microbial sources of such lipids because our knowledge of lipid composition in different microorganisms has been developed on the limited subset of species that have been cultivated (known to be a small percentage of extant microbial species; e.g., Ward et al., 1992), and usually without connection between the cultivated species analyzed and the habitats in which they might leave their lipid remnants (Des Marais et al., 1992a). Therefore, we focused our studies on hot spring microbial mat habitats and the species cultivated from them. It should be obvious from the preceding section that these species still constitute a small proportion of those that occur in such habitats, but at least they are from the same system, and at least the concurrent 16S rRNA work helps us measure the difference between the culture collections and the real natural biodiversity, which is minimized as new cultivation strategies are devised (Ferris et al., 1996b; Ward et al., 1997).

In past work, we demonstrated that the origins of neutral lipids and acyl constituents of complex polar lipids found in hot spring mats can be understood in terms of lipids known to be produced by bacteria isolated from these habitats (Dobson et al., 1988; Sheia et al., 1990, 1991; Ward et al., 1985, 1989a, 1994b; Zeng et al., 1992a,b). The relative abundances (Ward et al., 1989a) and vertical distributions of these lipids (Zeng et al., 1992b) suggested that the lipids reflect the trophic structure of mat communities. For example, lipids of cyanobacteria and green nonsulfur bacteria are more abundant than those of aerobic and anaerobic chemoorganotrophic bacteria, which are in turn more abundant than lipids of terminal anaerobes, such as sulfate reducers and methanogens. This is as expected given losses as energy is transferred from primary producers to organisms at higher trophic levels.

A major problem, however, is the limited power of lipid biomarkers to resolve microbial populations. A perfect illustration of this point is provided by the most abundant polar lipid-derived fatty acids (PLFAs) in the Octopus Spring mat ($C_{16:0}$, $C_{16:1}$, $C_{18:0}$, and

$C_{18:1}$) that are common to both of the major mat phototrophs, cyanobacteria and *Chloroflexus*. These PLFAs are very common among organisms in general and therefore nondiagnostic and also the inputs of the two types of organisms cannot be resolved. More recently, we explored the utility of direct analysis of intact complex lipids of the mat and two phototrophs using fast-atom-bombardment mass spectrometry (FABMS) (Ward et al., 1994b). Both cyanobacteria and *Chloroflexus* make mono- and diglycosyl diglycerides (MG and DG) and phosphatidyl glycerol (PG), but only cyanobacteria make sulfoquinosovyl diglycerides (SQ), which are therefore diagnostic. FABMS provided information to suggest the presence of three major mass clusters of complex lipids in the mat (Fig. 4a). Using thin layer chromatography (TLC), it was possible to separate and identify these lipid functional classes (Fig. 4b). By analyzing TLC-purified samples by FABMS, the specific masses measured in FABMS could be associated with their functional classes (shown for SQ in Fig. 4c). In this way, we were able to recognize that the major complex lipids of the mat, MG, DG, and SQ, most closely matched those of cyanobacteria. The specific masses of these complex lipids did not correspond exactly to the masses of lipids produced by the cyanobacterial culture *Synechococcus lividus* either due to environmental differences

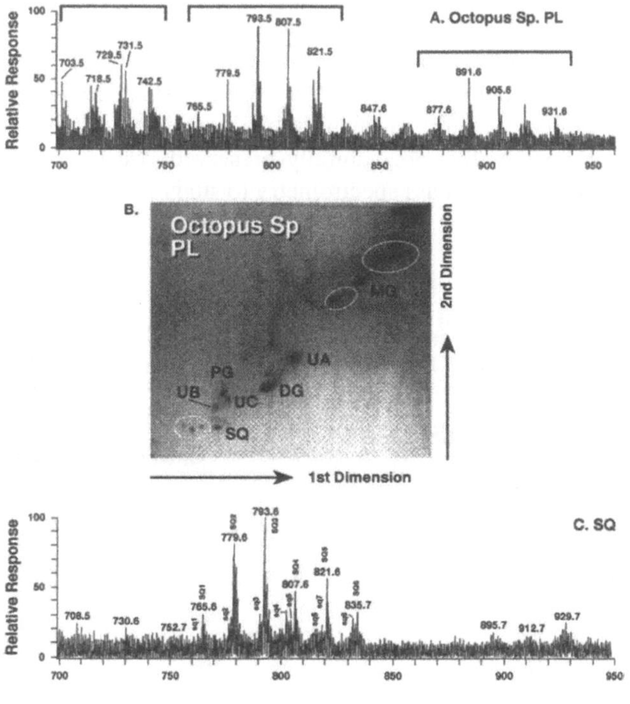

Figure 4. Analysis of complex polar lipids in the phospholipid (PL)-containing fraction extracted from the 50°C Octopus Spring cyanobacterial mat. (a) FABMS spectrum of the whole fraction, (b) TLC analysis of the whole fraction, and (c) FABMS spectrum of TLC-purified sulfoquinosovyl diglyceride (SQ) components. MG and DG, mono- and diglycosyl diglycerides; PG, phosphatidyl glycerol; UA and UB, unknowns; encircled compounds are pigments. Modified from Ward et al. (1994).

between nature and culture or to the fact that this cultivated species is not necessarily a good representative for the ten other cyanobacteria that inhabit the mat and are likely to be more abundant as judged by molecular and culture studies (Ferris et al., 1996a,b; Ferris and Ward, 1997; Ruff-Roberts et al., 1994; Ward et al., 1997).

4. NATURALLY PRESERVED BIOMARKERS CAN BE RELATED TO THEIR MICROBIAL SOURCES

For a cell component to enter the chemical fossil record, it must first survive decomposition within the mat system. In this regard, 16S rRNA, complex polar lipids, and their constituents all exhibit decreasing concentrations with depth in mats, as shown in Fig. 5. One way to circumvent this problem is to study biomarkers that occur in the chemical fossil record. For instance, mid-chain-branched monomethylalkanes exist in Precambrian oils (Des Marais et al., 1992a; Summons, 1992). By showing that such compounds can be correlated with modern mats that contain cyanobacteria (Ward et al., 1989a) and that only modern cyanobacteria produce these compounds (Shiea et al., 1990), we help associate the origin of these compounds with the possible inputs of cyanobacteria. Another class of compounds that occur in Precambrian sediments and oils is steranes, derived from sterols (Des Marais et al., 1992a). By showing that sterols are abundant only in mats containing eucaryotic species, such as algae and fungi, we help associate the origin of steranes with the possible inputs of eucaryotes (Shiea et al., 1991; Ward et al., 1989a).

Another way to circumvent the problem of biodegradation of biomarkers is to focus attention on cell components that are naturally preserved. We have begun to use flash pyrolysis–gas chromatography–mass spectrometry to study residual organic material that

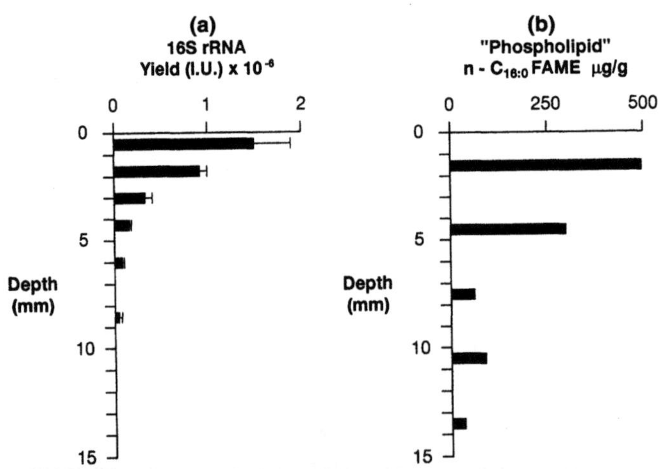

Figure 5. Vertical profiles of (a) 16S rRNA (data from Ruff-Roberts et al., 1994), and (b) hexadecanoic-acid fatty acid methyl esters (FAME) released by methanolysis from complex lipids in the phospholipid-containing fraction of ca. 3-mm depth intervals of a 50°C region of the Octopus Spring cyanobacterial mat (data from Zeng, 1988).

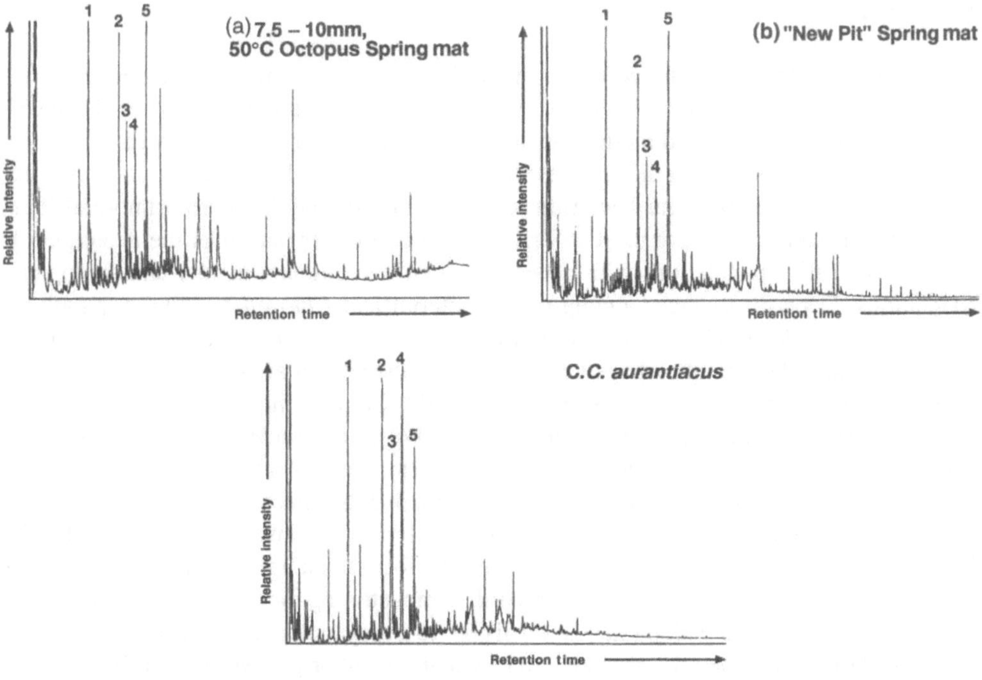

Figure 6. Flash pyrolysis-gas chromatography traces of the residue after lipid extraction (see Ward et al., 1994) and saponification of (a) the 7.5- to 10-mm depth interval of a 50°C Octopus Spring cyanobacterial mat sample, (b) the "New Pit" Spring *Chloroflexus* sp. mat, and (c) *Chloroflexus aurantiacus* strain Y-400-fl. Numbered peaks correspond to polysaccharide pyrolytic products identified by pyrolysis-gas chromatography-mass spectrometric analysis.

remains following lipid extraction and after removal of readily saponifiable compounds. By focusing attention on deep mat layers, we were able to detect materials that are naturally preserved (Fig. 6a). Pyrolysis–gas chromatography–mass spectrometry showed that some of the pyrolytic products were derived from polysaccharides, and through comparative pyrolysis–gas chromatographic analysis, we could associate these with products found in *C. aurantiacus* and the "New Pit" Spring mat, which is constructed through the photoautotrophic activity of *Chloroflexus* sp. (Fig. 6b and c). Survival of *Chloroflexus* remnants deep in mats is consistent with previous evidence of the persistence of *Chloroflexus*-specific antigens in deeper mat layers (Tayne et al., 1987). Of course, *Chloroflexus* populations also exist in Octopus Spring, but here the evidence suggests they are mainly photoheterotrophic and survive by cross-feeding organic compounds from cyanobacteria (Bateson and Ward, 1988; Anderson et al., 1987). One intriguing possibility that we are currently exploring is that these naturally preserved *Chloroflexus* biomarkers might record unique stable carbon isotopic signatures when the organism is growing photoautotrophically, because the pathway for CO_2 fixation in the more primordial green nonsulfur bacteria is different from that of the more recently evolved cyanobacteria, which use the Calvin cycle (Holo and Sirevåg, 1986; Ivanovsky et al., 1993; Strauss and Fuchs, 1993).

5. SUMMARY

Studies of hot spring microorganisms and microbial communities are relevant to microbial evolution because of the probable antiquity of thermophily and the existence of mat communities (considered modern analogs of stromatolites) that are formed by phototrophic microorganisms with different evolutionary histories. Most notably, mats that are formed by the green nonsulfur bacterium *Chloroflexus* sp. may be more primordial than cyanobacterial mats. We have used 16S rRNA as a biomarker that links the detection of many of the native populations with patterns of their evolutionary history and ecological adaptations. Deep divergences among cyanobacterial and *Chloroflexus*-like populations cannot yet be understood in ecological terms, but combined results indicate that populations of cyanobacteria (and perhaps *Chloroflexus* relatives) that are specialized to physiochemical parameters may have evolved (coevolved?) from recent common ancestors. To help interpret the organic geochemical record, we have tried to understand the connections between lipid biomarkers and their microbial sources; most recently we used FABMS to demonstrate that complex polar lipids and their components reflect the inputs of cyanobacteria, the predominant phototrophs in cyanobacterial mats. Because rRNA and most lipid biomarkers decompose rapidly upon burial, we have focused on lipid biomarkers that occur in the Precambrian organic geochemical record, demonstrating for instance that mid-chain-branched mono- and dimethylalkanes are likely to be cyanobacterial in origin and that steranes (derived from sterols) are likely to be eukaryotic in origin. Most recently, we have begun to explore naturally preserved organic components, and we discovered polysaccharide remnants and lipids likely to be from *Chloroflexus* sp., which might have the potential to distinguish whether this organism was the primary photoautotroph or a secondary photoheterotroph in the mat in which it grew. Results indicate that compound-specific stable carbon isotope analysis of lipids unique for cyanobacteria and *Chloroflexus* sp. can provide such information (Meer et al., 2000; Madigan, this volume).

ACKNOWLEDGMENTS. This work was funded by grants from the National Science Foundation (BSR-9209677) and the National Aeronautic and Space Administration (NAGW-2764) and gifts from Stratagene Corporation (LaJolla, CA). We also thank the U.S. National Park Service for its assistance in work at Yellowstone National Park. We are indebted to our other organic geochemistry collaborators, Dr. Geoffrey Eglinton (Organic Geochemistry Unit, Bristol University) and Dr. Herb Fredrickson (formerly of the Gesellschaft fur Biotechnologische Forschung, Braunschweig, Germany) and their associates, as well as to M. Baas (NIOZ) who joined us in the efforts reviewed herein. This is NIOZ publication number 3046.

REFERENCES

Anderson, K. L., Tayne, T. A., and Ward, D. M. 1987. Formation and fate of fermentation products in hot spring cyanobacterial mats. *Appl. Environ. Microbiol.* **53**:2343–2352.

Barns, S. M., Fundyga, R. E., Jeffries, M. W., and Pace, N. R. 1994. Remarkable Archaeal diversity detected in a Yellowstone National Park hot spring environment. *Proc. Natl. Acad. Sci. USA* **91**:1609–1613.

Barns, S. M., Delwiche, C. F., Palmer, J. D., and Pace, N. R. 1996. Perspectives on archaeal diversity, thermophily and monophyly from environmental rRNA sequences. *Proc. Natl. Acad. Sci. USA* **93**:9188–9193.

Bateson, M. M., and Ward, D. M. 1988. Photoexcretion and consumption of glycolate in a hot spring cyanobacterial mat. *Appl. Environ. Microbiol.* **54**:1738–1743.

Begon, M., Harper, J. L., and Townsend, C. R. 1990. *Ecology: Individuals, populations and communities*, 2nd ed. Cambridge, MA: Blackwell Scientific.

Bhattacharya, D., and Medlin, L. 1995. The phylogeny of plastids: A review based on comparisons of small-subunit ribosomal RNA coding regions. *J. Phycol.* **31**:489–498.

Brock, T. D. 1978. *Thermophilic microorganisms and life at high temperatures*. New York: Springer-Verlag.

Brock, T. D. 1987. The study of microorganisms *in situ*: Progress and problems. *Symp. Soc. Gen. Microbiol.* **41**: 1–17.

D'Amelio, E. D., Cohen, Y., and Des Marais, D. J. 1989. Comparative functional ultrastructure of two hypersaline submerged cyanobacterial mats: Guerrero Negro, Baja California Sur, Mexico, and Solar Lake, Sinai, Egypt. In Cohen, Y., and Rosenburg, E. (eds.), *Microbial mats: Physiological ecology of benthic microbial communities* (pp. 97–113). Washington, DC: American Society of Microbiology.

Des Marais, D. J., Bauld, J., Palmisano, A. C., Summons, R. E., and Ward, D. M. 1992a. The biogeochemistry of carbon in modern microbial mats. In Schopf, J. W., and Klein, C. (eds.), *The Proterozoic biosphere: A multidisciplinary study* (pp. 299–308). New York: Cambridge University Press.

Des Marais, D. J., D'Amelio, E., Farmer, J. D., Jorgensen, B. B., Palmisano, A. C., and Pierson, B. K. 1992b. Case study of a modern microbial mat-building community: The submerged cyanobacterial mats of Guerrero Negro, Baja California Sur, Mexico. In Schopf, J. W., and Klein, C. (eds.), *The Proterozoic biosphere: A multidisciplinary study* (pp. 325–333). New York: Cambridge University Press.

Dobson, G., Ward, D. M., Robinson, N. R., and Eglinton, G. 1988. Biogeochemistry of hot spring environments: Free lipids of a cyanobacterial mat. *Chem. Geol.* **68**:155–179.

Ferris, M. J., and Ward, D. M. 1997. Seasonal distributions of dominant 16S rRNA-defined populations in a hot spring microbial mat examined by denaturing gradient gel electrophoresis. *Appl. Environ. Microbiol.* **63**: 1375–1381.

Ferris, M. J., Muyzer, G., and Ward, D. M. 1996a. Denaturing gradient gel electrophoresis profiles of 16S rRNA-defined populations inhabiting a hot spring microbial mat community. *Appl. Environ. Microbiol.* **62**: 340–346.

Ferris, M. J., Ruff-Roberts, A. L., Kopczynski, E. D., Bateson, M. M., and Ward, D. M. 1996b. Enrichment culture and microscopy conceal diverse thermophilic *Synechococcus* populations in a single hot spring microbial mat habitat. *Appl. Environ. Microbiol.* **62**:1045–1050.

Ferris, M. J., Nold, S. C., Revsbech, N. P., and Ward, D. M. 1997. Population structure and physiological changes within a hot spring microbial mat community following disturbance. *Appl. Environ. Microbiol.* **63**:1367–1374.

Ferris, M. J., Nold, S. C., Santegoeds, C. M., and Ward, D. M. (this volume). Examining bacterial population diversity within the Octopus Spring microbial mat.

Giovannoni, S., Revsbech, N. P., Ward, D. M., and Castenholz, R. W. 1987. Obligately phototrophic *Chloroflexus*: Primary production in anaerobic hot spring microbial mats. *Arch. Microbiol.* **147**:80–87.

Giovannoni, S. J., Turner, S., Olsen, G. J., Barns, S., Lane, D. J., and Pace, N. R. 1988. Evolutionary relationships among cyanobacteria and green chloroplasts. *J. Bacteriol.* **170**:3584–3592.

Holo, H., and Sirevåg, R. 1986. Autotrophic growth and CO_2 fixation of *Chloroflexus aurantiacus*. *Arch. Microbiol.* **145**:173–180.

Ivanovsky, R. N., Krasilnikova, E. N., and Fal, Y. I. 1993. A pathway of the autotrophic CO_2 fixation in *Chloroflexus aurantiacus*. *Arch. Microbiol.* **159**:257–264.

Kasting, J. F., and Chang, S. 1992. Formation of the early earth and the origin of life. In Schopf, J. W., and Klein, C. (eds.), *The Proterozoic biosphere: A multidisciplinary study* (pp. 9–12). New York: Cambridge University Press.

Klein, C., and Beukes, N. J. 1992. Time distribution, stratigraphy, and sedimentologic setting, and geochemistry of Precambrian iron formations. In Schopf, J. W., and Klein, C. (eds.), *The Proterozoic biosphere: A multidisciplinary study* (pp. 139–146). New York: Cambridge University Press.

Meer, M. T. J., Schouten, S., de Leeuw, J. W., and Ward, D. M. 2000. Autotrophy of green nonsulphur bacteria in hot spring microbial mats: Biological explantions for isotopically heavy organic carbon in the geological record. *Env. Microbiol.* **2**:428–435.

Mojzsis, S. J., Arrhenius, G., McKeegan, K. D., Harrison, T. M., Nutman, A. P., and Friends, C. R. L. 1996. Evidence for life on Earth before 3,800 million years ago. *Nature* **384**:55–59.

Nold, S. C., and Ward, D. M. 1996. Photosynthate partitioning and fermentation in hot spring microbial mat communities. *Appl. Environ. Microbiol.* **62:**4598–4607.

Olsen, G. J., Woese, C. R., and Overbeek, R. 1994. The winds of (evolutionary) change: Breathing new life into microbiology. *J. Bacteriol.* **176:**1–6.

Pace, N. R. 1991. Origin of life—facing up to the physical setting. *Cell* **65:**531–533.

Pace, N. R. 1997. A molecular view of microbial diversity and the biosphere. *Science* **276:**734–740.

Pace, N. R., Stahl, D. A., Lane, D. J., and Olsen, J. G. 1986. The analysis of natural microbial populations by ribosomal RNA sequences. *Adv. Microbial Ecol.* **9:**1–55.

Peary, J. A., and Castenholz, R. W. 1964. Temperature strains of a thermophilic blue-green alga. *Nature* **202:** 720–721.

Pierson, B. K., and Castenholz, R. W. 1974. A phototrophic gliding filamentous bacterium of hot springs, *Chloroflexus aurantiacus*, gen. and sp. nov. *Arch. Microbiol.* **100:**5–24.

Pierson, B. K. 1992. Introduction. In Schopf, J. W., and Klein, C. (eds.), *The Proterozoic biosphere: A multidisciplinary study* (pp. 247–251). New York: Cambridge University Press.

Ramsing, N. B., Ferris, M. F., and Ward, D. M. 1997. Light-induced motility of thermophilic *Synechococcus* isolates from Octopus Spring, Yellowstone National Park. *Appl. Environ. Microbiol.* **63:**2347–2354.

Ramsing, N. B., Ferris, M. J., and Ward, D. M. 2000. Vertical positioning of microbial populations within the photic zone of a hot spring microbial mat community. *Env. Microbiol.* **66:**1038–1049.

Reysenbach, A.-L., Wickham, G. S., and Pace, N. R. 1994. Phylogenetic analysis of the hyperthermophilic pink filament community in Octopus Spring, Yellowstone National Park. *Appl. Environ. Microbiol.* **60:**2113–2119.

Ruff-Roberts, A. L., Kuenen, J. G., and Ward, D. M. 1994. Distribution of cultivated and uncultivated cyanobacteria and *Chloroflexus*-like bacteria in hot spring microbial mats. *Appl. Environ. Microbiol.* **60:**697–704.

Santegoeds, C. M., Nold, S. C., and Ward, D. M. 1996. Denaturing gradient gel electrophoresis used to monitor the enrichment culture of aerobic chemoorganotrophic bacteria from a hot spring cyanobacterial mat. *Appl. Environ. Microbiol.* **62:**3922–3928.

Schopf, J. W. 1992a. Paleobiology of the Archaean. In Schopf, J. W., and Klein, C. (eds.), *The Proterozoic biosphere: A multidisciplinary study* (pp. 25–39). New York: Cambridge University Press.

Schopft, J. W. 1992b. Proterozoic prokaryotes: Affinities, geologic distribution and evolutionary trends. In Schopf, J. W., and Klein, C. (eds.), *The Proterozoic biosphere: A multidisciplinary study* (pp. 195–218). New York: Cambridge University Press.

Shiea, J., Brassell, S., and Ward, D. M. 1990. Mid-chain branched mono- and dimethylalkanes in hot spring cyanobacterial mats: A direct biogenic source for branched alkanes in ancient sediments. *Org. Geochem.* **15:** 223–231.

Shiea, J., Brassel, S., and Ward, D. M. 1991. Comparative analysis of free lipids in hot spring cyanobacterial and anoxygenic photosynthetic bacterial mats. *Org. Geochem.* **17:**309–319.

Strauss, G., and Fuchs, G. 1993. Enzymes of a novel autotrophic CO_2 fixation pathway in the phototrophic bacterium *Chloroflexus aurantiacus*, the 3-hydroxypropionate cycle. *Eur. J. Biochem.* **215:**633–643.

Strauss, H., Des Marais, D. J., Hayes, J. M., and Summons, R. E. 1992a. Concentrations of organic carbon and maturities and elemental compositions of kerogens. In Schopf, J. W., and Klein, C. (eds.), *The Proterozoic biosphere: A multidisciplinary study* (pp. 95–99). New York: Cambridge University Press.

Strauss, H., Des Marais, D. J., Hayes, J. M., and Summons, R. E. 1992b. The carbon-isotopic record. In Schopf, J. W., and Klein, C. (eds.), *The Proterozoic biosphere: A multidisciplinary study* (pp. 117–127). New York: Cambridge University Press.

Summons, R. E. 1992. Abundance and composition of extractable organic matter. In Schopf, J. W., and Klein, C. (eds.), *The Proterozoic biosphere: A multidisciplinary study* (pp. 101–115). New York: Cambridge University Press.

Summons, R. E., and Hayes, J. M. 1992. Principles of molecular and isotopic biogeochemistry. In Schopf, J. W., and Klein, C. (eds.), *The Proterozoic biosphere: A multidisciplinary study* (pp. 83–93). New York: Cambridge University Press.

Tayne, T. A., Cutler, J. E., and Ward, D. M. 1987. Use of *Chloroflexus*-specific antiserum to evaluate filamentous bacteria of a hot spring microbial mat. *Appl. Environ. Microbiol.* **53:**1965–1968.

Walter, M. R., Grotzinger, J. P., and Schopf, J. W. 1992. Proterozoic stromatolites. In Schopf, J. W., and Klein, C. (eds.), *The Proterozoic biosphere: A multidisciplinary study* (pp. 253–260). New York: Cambridge University Press.

Ward, D. M., and Castenholz, R. W. In press. Cyanobacteria in geothermal habitats. In Potts, M., and Whitton, B. (eds.), *Ecology of Cyanobacteria*. Dordrecht, The Netherlands: Kluwer Academic.

Ward, D. M., Beck, E., Revsbech, N. P., Sandbeck, K. A., and Winfrey, M. R. 1984. Decomposition of hot spring microbial mats. In Cohen, Y., Castenholz, R. W., and Halvorson, H. O. (eds.), *Microbial mats: Stromatolites* (pp. 191–214). New York: A. R. Liss.

Ward, D. M., Brassell, S. C., and Eglinton, G. 1985. Archaebacterial lipids in hot-spring microbial mats. *Nature* **318**:656–659.

Ward, D. M., Tayne, T. A., Anderson, K. L., and Bateson, M. M. 1987. Community structure and interactions among community members in hot spring cyanobacterial mats. *Symp. Soc. Gen. Microbiol.* **41**:179–210.

Ward, D. M., Weller, R., Shiea, J., Castenholz, R. W., and Cohen, Y. 1989a. Hot spring microbial mats: Anoxygenic and oxygenic mats of possible evolutionary significance. In Cohen, Y., and Rosenburg, E. (eds.), *Microbial mats: Physiological ecology of benthic microbial communities* (pp. 3–15). Washington, DC: American Society of Microbiology.

Ward, D. M., Shiea, J., Zeng, Y. B., Dobson, G., Brassell, S., and Eglinton, G. 1989b. Lipid biochemical markers and the composition of microbial mats. In Cohen, Y., and Rosenburg, E. (eds.), *Microbial mats: Physiological ecology of benthic microbial communities* (pp. 439–454). Washington, DC: American Society of Microbiology.

Ward, D. M., Weller, R., and Bateson, M. M. 1990. 16S rRNA sequences reveal numerous uncultured microorganisms in a natural community. *Nature* **345**:63–65.

Ward, D. M., Bateson, M. M., Weller, R., and Ruff-Roberts, A. L. 1992. Ribosomal analysis of microorganisms as they occur in nature. *Adv. Microb. Ecol.* **12**:219–286.

Ward, D. M., Ferris, M. J., Nold, S. C., Bateson, M. M., Kopczynski, E. D., and Ruff-Roberts, A. L. 1994a. Species diversity in hot spring microbial mats as revealed by both molecular and enrichment culture approaches—relationship between biodiversity and community structure. In Stal, L. J., and Coumette, P. (eds.), *Microbial mats: Structure, development and environmental significance* (pp. 33–44, Series G: Ecological Sciences vol. 35, NATO/ASI Series). Heidelburg: Springer-Verlag.

Ward, D. M., Panke, S., Kloeppel, K. D., Christ, R., and Fredrickson, H. 1994b. Complex polar lipids of a hot spring cyanobacterial mat and its cultivated inhabitants. *Appl. Environ. Microbiol.* **60**:3358–3367.

Ward, D. M., Santegoeds, C. M., Nold, S. C., Ramsing, N. B., Ferris, M. J., and Bateson, M. M. 1997. Biodiversity within hot spring microbial mat communities: Molecular monitoring of enrichment cultures. *Antonie Leeuwenhoek* **71**:143–150.

Ward, D. M., Bateson, M. M., Ferris, M. J., and Nold, S. C. 2000. Biodiversity, evolution and ecology of microorganisms inhabiting hot spring microbial mat communities. *Microbiol. Mol. Biol. Rev.* **62**:1353–1370.

Weller, R., Bateson, M. M., Heimbuch, B. K., Kopczynski, E. D., and Ward, D. M. 1992. Uncultivated cyanobacteria, *Chloroflexus*-like inhabitants, and spirochete-like inhabitants of a hot spring microbial mat. *Appl. Environ. Microbiol.* **58**:3964–3969.

Widdel, F., Schnell, S., Heising, S., Ehrenreich, A., Assmus, B., and Schink, B. 1993. Ferrous iron oxidation by anoxygenic phototrophic bacteria. *Nature* **362**:834–835.

Woese, C. R. 1987. Bacterial evolution. *Microbiol. Rev.* **51**:221–271.

Woese, C. R. 1994. There must be a prokaryote somewhere: Microbiology's search for itself. *Microbiol. Rev.* **58**: 1–9.

Zeng, Y. B. 1988. Geochemical studies of microbial lipids. Ph.D. thesis. University of Bristol, Bristol, UK.

Zeng, Y. B., Ward, D. M., Brassell, S., and Eglinton, G. 1992a. Biogeochemistry of hot spring environments. 2. Lipid compositions of Yellowstone (Wyoming, U.S.A.) cyanobacterial and *Chloroflexus* mats. *Chem. Geol.* **95**:327–345.

Zeng, Y. B., Ward, D. M., Brassell, S., and Eglinton, G. 1992b. Biogeochemistry of hot spring environments. 3. Apolar and polar lipids in the biologically active layers of a cyanobacterial mat. *Chem. Geol.* **95**:347–360.

13

Research Accomplishments of a Small Business Using Yellowstone's Extremophiles

Joan Combie and Kenneth Runnion

1. INTRODUCTION

It has been found that as conditions become more demanding in extreme environments, microorganisms predominate and certain microorganisms occupy the more extreme niches (Brock, 1986). Geothermal environments are among the most productive places to search for thermophiles. The Yellowstone ecosystem is the world's foremost showcase of geothermal activity. With nearly 10,000 hot springs and steam vents, Yellowstone has the largest and most varied array of accessible, geothermal phenomena on earth. Nowhere else is such a diversity of extreme habitats found. Water temperatures range from cold mountain streams to boiling geysers. Strongly acidic (pH below 1.0) and strongly basic (pH 10.5) waters are in this region. High concentrations of some ions and heavy metals present unique environments for evolutionary adaptations. Extreme microbial habitats include oligotrophic conditions, rich petroleum seeps, and rare, high temperature microbial mat communities. As temperatures increase, oxygen solubility decreases, resulting in numerous anaerobic habitats. At the opposite extreme are certain mat communities where oxygen production results in microhabitats supersaturated with respect to oxygen. It should be emphasized that organisms are not just exposed only to one of these extremes but generally to several extremes simultaneously. These habitats are relatively undisturbed, allowing long term evolution. It appears that this ecosystem provides an unparalleled opportunity to search for novel organisms.

Joan Combie and Kenneth Runnion • Montana Biotech, Belgrade, Montana 59714.

Thermophiles: Biodiversity, Ecology, and Evolution, edited by Reysenbach *et al.* Kluwer Academic/Plenum Publishers, New York, 2001.

2. RESULTS AND DISCUSSION

2.1. Diversity of Collection and Isolation Techniques

To avoid recovering the same handful of organisms over and over, a diversity of collection and isolation techniques were used.

The material collected included samples from the water column, sediment and scrapings off rock walls of pools, and small pieces of microbial mats. During transportation to the laboratory, some samples were held at the temperature of the original habitat by placing tubes in thermoses filled with water from the collection site. In other cases, samples were frozen on dry ice or processed and added to initial incubation media in the field. Some samples were held at elevated temperatures to eliminate lower temperature contaminants, and some were treated with antimicrobial agents. Serial dilutions were particularly effective for recovering slow growing but numerically dominant species (Combie and Runnion, 1996).

The nature of the medium was one of the easiest factors to manipulate. On occasion, when the chemical analysis of a collection site was known, media were designed on the basis of these data. More frequently, several different media were used for recovering organisms from a given site. To select organisms that could use a given substrate, that compound was added to the medium. For example, when searching for microorganisms to remove polyurethane paint, water samples were incubated with minimal salts and polyurethane paint. Only organisms that could use the paint as a source of carbon grew. Thus a costly screening step was eliminated, leaving only organisms that had a reasonable possibility of some activity on paint (Combie and Runnion, 1996).

2.2. Heat-Stable Enzymes

Although enzymes can be stabilized by chemical and physical methods, these tend to be expensive. Alternatively, a stable form of the desired enzyme may be obtained from microorganisms that inhabit extreme environments, particularly those at high temperatures. Amine oxidase, glucose oxidase, peroxidase, urease, and alkaline phosphatase were among the thermally stable enzymes found. Research results for three enzymes are presented here.

2.2.1. Peroxidase

Peroxidase, an enzyme of the oxygen defense system, is frequently used as a label in ELISA type assays. Poor temperature stability of the widely available horseradish peroxidase, led to a search for a more thermally tolerant source. Several peroxidase-producing microorganisms were isolated, including both acid and near neutral forms of peroxidase. One isolate was from a habitat where pH was 2.4 and the temperature 47°C. Not surprisingly, the resulting peroxidase was most active and most stable under strongly acidic conditions. In stability tests, nearly complete activity was retained for two months at 45°C and a pH of 2–3. At 50°C, the half-life was one week; at 70°C, only 8% of the activity remained after 24 hours. The half-life for horseradish peroxidase was 24 hours at 45°C (Combie et al., 1992).

The original habitat for another peroxidase-producing isolate had a temperature of 51°C and pH of 6.5. The optimal pH for this peroxidase was 5.0 with citric acid or HEPES buffer and 6.5 with Tris base. At 60°C, the half-life was 48 hours. The enzyme was active between 20 and 50°C. At temperatures above 50°C, the activity dropped off possibly due to instability of the substrate, hydrogen peroxide. The 16S rRNA sequence of this organism identified it as a novel isolate (Combie et al., 1992).

2.2.2. Urease

Novel urease-based immunoassays using potentiometry with a silicon sensor and fluorescence with fiber-optic technology have been developed. Urease catalyzes the hydrolysis of urea to form eventually two molecules of ammonia. At near neutral pH, ammonia molecules become protonated, resulting in a net pH increase, a reaction easily followed by pH sensors. The substrate, urea, is safe, inexpensive, and stable in solution. The absence of urease in most mammalian tissues reduces background noise and results in increased sensitivity in assays that involve mammalian cells.

Urease is found in many plants, bacteria, fungi, yeast, and algae. Commercially available urease comes largely from jack beans, but it must be stored at 4°C or lower for stability. Although thermophiles would be an obvious potential source of a more stable urease, the only definitive reference to such an organism, when this research was initiated, was in a Japanese patent on *Bacillus* sp. TB-90 (Takashio et al., 1988).

Two hundred and thirty cultures from Yellowstone thermal waters were tested for urease production. Only twenty-seven isolates yielded a positive reaction in the screening test; many were eliminated following more definitive testing. Partially purified heat stable urease that had an optimum initial pH of 5.5–7.0 in 50 mM citrate buffer was obtained. The enzyme was active over a wide temperature range, crucial for field applications. Stability analysis at 50°C for 209 hours revealed 94% activity retention compared to 7% retention for jack bean urease. Ureases from diverse sources vary widely in molecular weight but are generally quite large. This poses an obstacle to some applications such as those where steric hindrance becomes a problem. For example, jack bean urease has a molecular weight of 550,000, and most bacterial ureases range from 200,000 to 380,000. One heat stable urease found in Yellowstone had a molecular weight of 125,000 (Combie et al., 1992; Runnion et al., 1992).

2.2.3. Alkaline Phosphatase

Alkaline phosphatase has been used in a number of biosensors. Coupling an alkaline phosphatase substrate to a redox compound allows using certain types of potentiometric sensors. An amperometric biosensor that uses (*N*-ferrocenoyl)-4-aminophenol phosphate as the alkaline phosphatase substrate has been developed. Thermophilic microorganisms that produce readily detectable levels of alkaline phosphatase were isolated. One isolate produced a very stable alkaline phosphatase (Table 1). Because the enzyme was excreted, the purification process was simplified (Combie et al., 1994).

2.3. Carotenoids

Carotenoids enjoy a rapidly growing market where these natural products can replace chemically produced alternatives. Certain carotenoids are already included in multivitamin preparations as antioxidants. Others replace colorants in food or are added to feed for

Table 1
Thermal Stability of Alkaline Phosphatase[a]

Boiling time (min)	% Activity remaining
0	100
15	103
30	99
45	98
60	98

[a]Partially purified enzyme was boiled in 20 mM Tris, pH 8.5, 10 mM calcium chloride, and 1 mM magnesium chloride.

aquacultured fish and shrimp. The natural products are preferable toxicologically, and they are also often less expensive. In microorganisms, carotenoids provide a protective effect resulting from their ability to react with and quench toxic oxygen species. While searching thermal waters for microorganisms that produce enzymes of the oxygen defense system, it was noticed that organisms that have high enzyme activity were unpigmented despite the fact that collections were from pigmented microbial mat communities. These laminated mats are subjected to intense solar radiation and are often well over oxygen saturation due to the activity of phototrophs. The fact that pigmented forms lacked the enzyme defense mechanisms suggested that carotenoids provide extremely effective protection. A collection of pigmented organisms resulted in recovering bacteria that produce carotenoids of commercial interest.

2.4. TNT Degradation

Nitro-substituted compounds such as nitrobenzene and TNT are difficult to remove from contaminated soil and waste streams. Biotreatment would be less expensive than incineration, but some attempts have resulted in accumulation of toxic intermediates. Microorganisms from extreme environments are a rich source of unique metabolic pathways. Based on success with anaerobes in previous related work and the development of a TNT degrading consortium, thermophilic anaerobes were the focus of a preliminary study. Using enrichment cultures to select organisms that had the greatest potential, anaerobic thermophiles were tested for the ability to degrade nitro-substituted munitions. Preliminary results are shown in Table 2. Samples of the liquid and soil were tested for the parent compound and metabolites. Much of the TNT had disappeared; the major metabolite identified was 4-amino-2,6-dinitrotoluene.

2.5. Coal Biosolubilization

Biological processing of coal at mild operating conditions could provide significant economic and environmental advantages compared with thermal/chemical coal conversion processes. Water samples were collected from Yellowstone. Thermophiles were isolated and screened for their ability to solubilize coal. Leonardite, a naturally weathered lignite

Table 2
Biodegradation of TNT

Sample tested	TNT (ppm)	4-Amino-2,6-dinitrotoluene (ppm)
Starting material	1200.0	—
Liquid	41.4	2.5
Soil extract	50.0	1.1

from North Dakota, was used for screening tests. The microorganisms could function over a wide pH range, act rapidly because of higher operating temperatures, and survive high levels of toxic metal commonly associated with coal. Isolates that had the highest biosolubilization activity came from habitats where pH was between 6 and 9 and where temperatures were between 45 and 65°C. Cultures above pH 9 were difficult to grow on the media needed for biosolubilization. Below pH 6, isolates showed minimal or no solubilizing activity (Runnion and Combie, 1990).

Viable thermophiles were recovered from liquid cultures 12 days after coal had been added. A medium that buffered the culture as coal went into solution was important. Cultures and controls showed a direct correlation between coal biosolubility and pH. The mechanisms of solubilization seemed to be alkali-mediated. No enzymatic mediated solubilization was detected. Metal ion sequestering agents (chelators) contributed to solubilization (Runnion and Combie, 1990).

2.6. Microbial Coal Desulfurization

The Clean Air Act dictated that annual SO_2 emissions from coal-fired power plants be decreased by ten million tons by the year 2000. Microbial desulfurization can lower capital and operating costs, does not reduce the coal heating value, and can remove a higher portion of the sulfur compared with existing chemical and physical cleaning methods. Sulfur occurs in coal in two forms, inorganic and organic. It has been established that autotrophic bacteria, including *Thiobacillus* and *Sulfolobus* can decompose pyrite, the major inorganic sulfur component. Removal of the organic sulfur fraction has proven far more difficult. Hot sulfurous waters at neutral pH were relatively rare in Yellowstone compared with the abundant number of strongly acidic waters that had high levels of sulfur compounds. While research has been concentrated on acidophilic autotrophs, these neutral waters were left largely unstudied with respect to coal desulfurization.

Dibenzothiophene (DBT) was used as an organic sulfur model compound to screen for organisms that have the desired metabolic routes. Cultures that can metabolize DBT by the sulfur-specific pathway fluoresce purple when examined under short wavelength UV light. Water samples were collected from sulfurous, near neutral pH, thermal waters of Yellowstone. Fifty-nine microbial communities were screened for DBT desulfurization using the UV-fluorescence assay. The eighteen positive cultures were tested for coal desulfurization using the North Dakota lignite. One thermophilic culture was incubated for 30 days in a 5-liter reactor with Leonardite. More than 90% pyritic sulfur, 33% organic sulfur, and 50% total sulfur were removed. Key to this success was the fact that the organisms thrived in the presence of high levels of metal in coal and under other environmental extremes (Runnion and Combie, 1993).

2.7. Polyurethane Paint Removal

Traditional methods for paint removal, chemical stripping and sandblasting, generate hazardous waste materials. Under the Small Business Innovation Research Program (SBIR), the U.S. Navy funded a project to look for alternatives to remove polyurethane aircraft coatings (Runnion and Combie, 1995).

Thermophilic microorganisms were isolated from the thermal waters of Yellowstone and screened for their ability to remove polyurethane coatings from test coupons. A total of 157 cultures were incubated with test coupons for 72 hours. Forty-one were positive in two rounds of testing. Tests involved the removal of polyurethane coatings from aluminum coupons by the tape test. Cell free extracts were also tested and produced significant blistering of the coupons in 1 to 3 days. It was determined that the activity was exported into the medium, simplifying the purification process and avoiding the inclusion of whole cells in the depainting preparation.

Cells and cell free extracts were assayed for urease, amine oxidase, esterase, and protease. None of these enzymes individually correlated with polyurethane removal. Currently, it is believed a combination of enzymes and/or other microbially produced compounds are responsible for the polyurethane removal activity (Runnion and Combie, 1995). A variety of surface analysis techniques were used to study the mechanism of action. Polyurethane removal is probably a multistep process that requires several enzymes and possibly certain metabolites or cofactors. The mechanism of blistering involved swelling of the polyurethane topcoat, migration of enzymes to the aluminum substrate surface, and disruption of the adhesion at the epoxy/aluminum interface (Runnion and Combie, 1995).

Six cultures were characterized by 16S rRNA sequence analysis. The microorganisms fell into different groups of the genera *Bacillus* and *Paenibacillus*.

3. CONCLUSIONS

We have been collecting microorganisms from Yellowstone National Park for more than 11 years in search of novel organisms that have commercial applications. The 10,000 thermal features provide an unparalleled variety of extreme microbial habitats. Research has emphasized thermally stable enzymes, carotenoids, and bioprocessing. Enzymes identified include amine oxidase, glucose oxidase, peroxidase, urease, and alkaline phosphatase. Carotenoids for use as antioxidants, food colorants, and feed additives have been recovered from previously unknown bacteria. Bioprocessing projects have included TNT degradation, microbial coal solubilization, biological coal desulfurization, and enzymatic polyurethane paint removal.

ACKNOWLEDGMENTS. This work was supported by SBIR grants from the DOD, DOE, and EPA.

REFERENCES

Brock, T. D. (ed.). 1986. *Thermophiles*. New York: Wiley.

Combie, J., and Runnion, K. 1996. Diversity in Yellowstone extremophiles. *J. Indust. Microbiol.* **17**:214–218.

Combie, J., Runnion, K., and Williamson, M. 1992. Acquisition of heat stable enzymes from thermophilic microorganisms: Peroxidases, ureases, and glucose oxidases, CRDEC-CR-152. Aberdeen Proving Ground, MD: Chemical Research, Development and Engineering Center.

Combie, J., Runnion, K., and Williamson, M. 1994. Heat stable alkaline phosphatase from thermophiles, CRDEC-CR-131. Aberdeen Proving Ground, MD: Chemical Research, Development and Engineering Center.

Runnion, K., and Combie, J. 1990. Thermophilic microorganisms for coal biosolubilization. *Appl. Biochem. Biotechnol.* **24/25**:1–928.

Runnion, K., and Combie, J. 1993. Organic sulfur removal from coal by microorganisms from extreme environments. *FEMS Microbiol. Lett.* **11**:139–144.

Runnion, K., and Combie, J. 1995. Microorganisms from extreme environments as source of thermally stable enzymes for removal of polyurethane aircraft coatings. Proceedings of the 1995 DOD/Industry Advanced Coatings Removal Conference, Columbus.

Runnion, K., Combie, J., and Williamson, M. 1992. Thermally stable urease from thermophilic bacteria. In Adams, M., and Kelly, R. (eds.), *Biocatalysis at extreme temperatures enzyme systems near and above 100°C*. Washington, DC: American Chemical Society.

Takashio, M., Yoneda, Y., and Mitani, Y. 1988. U.S. Patent 4,753,882.

14

The Yellowstone Microbiology Program
Status and Prospects

John D. Varley, Robert F. Lindstrom, and Charles C. Chester

1. INTRODUCTION

The U.S. National Park System (NPS) preserves and protects the nation's natural and cultural resources for public benefit. These resources provide unique opportunities for recreation and education, and also for advancing scientific endeavor. Archeology, botany, ecology, environmental monitoring, history, paleontology, zoology—the variety of research conducted in the national parks is extensive and, in places, intensive. Hosting over 200 research projects at any given time, Yellowstone National Park (YNP) indicates the general importance of National Parks for science. Yellowstone is particularly important, however, for the specific field microbiology. The Park's thermal features (e.g., hot springs, geysers, and fumaroles) harbor microorganisms that withstand extreme conditions of heat and acidity. Believed to be the world's greatest concentration of thermophilic biological diversity, Yellowstone is a strategic repository of unique genetic material.

Scientists first recognized the presence of microbial life in the Park's thermal features before 1900. But it was not until 1965, when Professor Thomas Brock began an extensive

John D. Varley and Robert F. Lindstrom • Yellowstone Center for Resources, Yellowstone National Park, Wyoming, 82190. **Charles C. Chester** • World Foundation for Environment and Development, Washington, D.C., 20036.

Thermophiles: Biodiversity, Ecology, and Evolution, edited by Reysenbach *et al.* Kluwer Academic / Plenum Publishers, New York, 2001.

study of Yellowstone's thermophilic microorganisms, that microbiologists began to fully appreciate the biological importance of Yellowstone's geothermal ecosystem. Yellowstone preserves geothermal habitats similar to those of the Earth's earliest biosphere, the environment in which DNA-based life first appeared. And the geochemical complexity of Yellowstone's estimated 10,000 individual thermal features maintains perhaps the most diverse concentration of geothermal habitat on the planet, where surficial temperatures are up to 94°C and pH levels from 0 to over 10 (Brock, 1978). In turn, this habitat diversity harbors beneficial conditions for a vast spectrum of thermophilic microorganisms.

From the development of phylogenetic analysis of DNA, where individual species (or "phylotypes" or "strain types") can be determined without the need to culture specimens, microbiologists are gaining insight into the true extent of the microbial world. Of the species present in Yellowstone's thermal features, researchers have concluded that less than 1% have been cultured (Pace et al., 1986). For example, enrichment culture shows the presence of two species of the genera *Synechococcus* and *Chloroflexus* in the Octopus Spring microbial mat. Phylogenetic analysis, however, has revealed a guild structure of dozens of interdependent species (Ward et al., 1994). Extrapolating the Octopus Spring data to the myriad thermal features throughout Yellowstone, it can be roughly estimated that hundreds, if not thousands, of undiscovered types of thermophilic microorganisms are present throughout the ecosystem.

Until recently, the value of Yellowstone's thermophilic microbial resources remained unknown, preserved inadvertently along with the "geological curiosities" for which the Park was established in 1872. But now microbiologists recognize that thermophiles produce uncommon, heat-stable enzymes and macromolecules. Because much of modern biotechnology is based on the use of enzyme catalysis for biochemical reactions, including genetic engineering, fermentation, and bioproduction of antibiotics, these heat-stable enzymes are becoming increasingly important in advancing science, medicine, and industry (Madigan et al., 1997). Such genetic information has already sparked a biotechnological revolution. In the late 1960s, Professor Brock discovered the thermophilic bacterium *Thermus aquaticus* in Yellowstone's Mushroom Pool (Brock and Freeze, 1969). About two decades later, biotechnologists incorporated the thermostable "*Taq*" polymerase enzyme from *T. aquaticus* into the polymerase chain reaction (PCR), a Nobel prize-winning invention that allows scientists to "amplify" minute samples of DNA (Gelfand et al., 1988). The use of *Taq* polymerase significantly enhanced PCR, enabling scientists to identify genetic material in practical applications ranging from forensic sciences and medical diagnostics to wildlife population studies and analysis of microbial diversity.

Yellowstone's microbial resources have received relatively little attention, but the Yellowstone Center for Resources (YCR) has been examining the subject for several years. To better manage Yellowstone's microbial resources, YCR is currently developing a Yellowstone Microbiology Program (YMP) that will address the following six programmatic areas: inventory and monitoring; research support; protection of geothermal habitat; benefit-sharing/bioprospecting; and education. Development of these six areas will improve our understanding of the Park's resources and our ability to protect those resources. For each of the above areas, this article examines (1) the current status of activity and (2) the prospects for development and improvement.

2. INVENTORY AND MONITORING OF YNP MICROORGANISMS

2.1. Current Status

An important component of the NPS mission is to inventory the biological resources of the parks (National Research Council, 1992; U.S. National Park Service, 1993; U.S. National Park Service, 1995). YCR has developed and currently maintains the *Thermophilic Microorganism Survey* (TMS) (Lindstrom, 1995), a baseline inventory of Yellowstone thermophiles that currently lists more than thirty microbial species found in Yellowstone. The inventory process is straightforward. Researchers funded by public and private organizations conduct field work (including specimen collection) in Yellowstone. As required by the NPS (36 CFR Ch.1 §2.5(g)(2)), published discoveries of new species are filed with the YCR during the annual investigators' reporting cycle. The information is processed into the TMS database, and the data are finally distributed to interested individuals.

For each species, the TMS provides eleven categories of information: scientific name, description, habitat type, YNP location, abundance, first observed, other locations, biotech use, citation, photographs (in YCR files), and specimens (status). Appendix 1 contains a sample record for *Thermus aquaticus*. The current TMS is an initial and informal survey and is not comprehensive.

2.2. Prospects

Due to (1) budgetary constraints throughout the entire NPS and (2) the significant taxonomic and database complexity of inventorying microorganisms, YCR has sought help from outside collaborators to further develop the TMS. Several independent scientists working in Yellowstone (from government, academia, and the private sector) are currently assisting YCR in developing the TMS. To understand, manage, and protect Yellowstone's vast microbial diversity, the Park needs an efficient method to collect, process, store, and disseminate an immense amount of data. The TMS could be improved in the following areas. The TMS should be relevant to resource management needs and therefore should be useful and accessible to Park managers in their efforts to protect Yellowstone's microbial resources. Access to the database should be open and convenient to various users in academia, other areas of the government, and private enterprise. The TMS should be made available via the Internet/World Wide Web (www). Additionally, YCR and collaborating scientists should publicize the significance and utility of the data, thereby potentially increasing scientific interest in Yellowstone's microbial resources. The database structure of the TMS should be compatible with Yellowstone's park-wide efforts to inventory its biological, geological, cultural, and other resources, including the Geographic Information System (GIS) database, which is currently under development by the YNP Spatial Analysis Center. Management and protocols for routine input of data into the TMS should be carefully considered to avoid duplication of efforts by coordination and consensus building among managers and scientists. Park managers and scientists will need to agree when research data should be incorporated into the TMS to ensure that the inventory is comprehensive and accurate and that entries are made in a timely fashion. A fundamental problem of microbiology has been disagreement over the classification of microorganisms into "species," "phylotypes," or "strain types." Hopefully, recent techniques of molecular phylogeny will allow microbiologists to quantitatively describe and differentiate between microorganisms. This is likely to form the basis for further development of the TMS.

3. MICROBIOLOGICAL RESEARCH SUPPORT

3.1. Current Status

The NPS has only a limited capacity to conduct state-of-the-art microbiological research. Consequently, the NPS promotes independent research in the national parks because the knowledge garnered adds to its understanding of and ability to manage park resources. Yellowstone supports qualified scientists by granting research permits without charging any processing costs and by waiving the Park's standard visitor entrance fee. Currently, YNP has granted research permits for more than 40 microbiological research projects. Park resource managers also accompany microbiologists into the field when an "official" or "uniformed" presence is necessary for safety and preservation (particularly during high public visibility research) and to document commercial collections. In such cases, the benefits are threefold: (1) the public is informed of the scientists' special status by the accompanying Park official, (2) the public is educated about the potential benefits of the scientists' activities, and (3) the scientists are freed from having to explain or defend their activities.

Perhaps the most important role of NPS staff is as an informational resource for visiting researchers. Park managers provide researchers with an intimate understanding and knowledge of the Park and its resources. Personal assistance from YNP officials will always be important, but researchers would undoubtedly benefit from more quantitative information about YNP's resources, particularly if that information were accessible outside of the Park through the Internet. To fill this gap, YNP's Spatial Analysis Center is currently developing a Geographic Information System for the Park which will be available in the future through the World Wide Web. The GIS database currently contains three "layers" that relate to thermal areas: a layer inventorying polygonal "hot" areas divided into three chemical types (acid, neutral, travertine), a layer inventorying geothermally affected soils (includes soils hydrothermally affected in the past or currently "hot"), and a layer inventorying geothermally influenced landforms.

3.2. Prospects

The YNP Spatial Analysis Center is currently planning several new layers for the GIS. These will include detailed base maps of thermal areas made from remotely sensed data, detailed thermal gradient maps, and a "point database" of named thermal features (that will include their geochemical characteristics). This information will be meshed with a "master geothermal ecosystem database" that will be established in conjunction with the Idaho National Engineering and Environmental Labs (INEEL). This information will be made available through the www.

4. PROTECTION OF GEOTHERMAL HABITAT

4.1. Current Status

Some concern has been raised over potential damage to the thermal features from microbial specimen collection and ecological experiments. Current regulations and permit agreements take this into account. According to the regulations, research that requires specimen collection is not permitted "if removal of the specimen would result in damage to their natural or cultural resources, affect adversely environmental or scenic values, or if the specimen is readily available outside of the park area" (36 CFR §2.5(b)). Additionally, when applying for a research permit at Yellowstone, researchers sign a form that states that "collecting shall be conducted in such a manner as not ... to cause damage to the environment." These rules, of course, do not guarantee that researchers will comply either with the letter or spirit of the regulations. In some cases, Park officials directly monitor and control collecting activities by accompanying microbiologists into the field. Due to staffing limitations, however, most research occurs without the presence of park officials.

4.2. Prospects

Ideally, all research conducted in the Park (not only microbiological research) would be directly accompanied and monitored by Park officials. This would provide direct protection for Park resources and also would directly inform Park officials of what researchers are learning during the course of their field investigations. However, even significant increases in staffing and budget are unlikely to allow Park officials to accompany all researchers in the Park. Additionally, most researchers are probably extremely careful to protect the resources they are studying, but it would be advisable for YNP to inform all researchers of explicit responsibilities. A list of these responsibilities could be sent to researchers along with their approved research permits, and could include the most basic obligations to practices that might not occur to researchers on their own. Accordingly, YCR is currently drafting a specific code of conduct, particularly for researchers who work in or near thermal features (e.g., microbiologists, geologists, geochemists, thermophilic plant botanists, etc.).

5. BENEFIT-SHARING/BIOPROSPECTING

5.2. Current Status

Microbiological research at Yellowstone has led to the development of at least one significant commercial product (*Taq* polymerase). Other potential products are currently under investigation. Several individuals within and outside the NPS have argued that Yellowstone should receive some of the commercial benefits accruing from research originating in the Park (Appendix II lists various articles on the issue of bioprospecting in YNP).

In 1994, the YCR staff began investigating the possibility of expanding the permitting process to allow the Park to sign a "research contract" with microbiologists who conduct research in the Park. The proposal would have earmarked a portion of commercial benefits derived from YNP microbes for the protection of Park resources. This general idea was widely debated and discussed among Park officials, academic and industry scientists, and conservationists during the 1995 Conference, *Biodiversity, Ecology, and Evolution of Thermophiles in Yellowstone National Park.* Although the participants reached no clear consensus on the issues of access to and use of YNP's thermophilic microorganisms, there was broad agreement that the thermophilic microbial resources should receive strong protection through Park–academia–industry collaboration.

Following the Conference, in January 1996 a "Blue Ribbon Panel" of Park officials, industry representatives, academic scientists, journalists, and conservationists attended a workshop in Costa Rica on the conservation and sustainable use of thermophilic biological resources. The purpose of this initiative was twofold. First, the workshop was an opportunity for divergent parties interested in YNP microorganisms to discuss and to evaluate management options in a neutral setting. The aim of the workshop was to examine how the use of microbial resources could benefit conservation, science, and economic development. The workshop did not result in policy recommendations, but participants agreed on the following principal issues and problems:

- emphasizing the importance of YNP thermophiles to science-based management throughout the U.S. National Park System;
- interpreting the legal and regulatory framework for the use of Park microorganisms by academia and the biotechnology industry;
- determining the "value" of thermophilic microbial diversity and its various beneficial uses;
- defining the role of the private sector in public conservation efforts;
- answering whether and how users of thermophilic microorganisms should compensate YNP or the National Park Service (NPS).

Second, the workshop was intended to introduce participants to the innovative approaches of the National Biodiversity Institute (INBio) of Costa Rica to the conservation and sustainable use of biodiversity. Participants learned about the INBio's four principal programs: (1) an inventory of the country's biodiversity, (2) a database for biodiversity information management, (3) a program for biodiversity information dissemination and education, and (4) a program for biodiversity prospecting (or "bioprospecting") in the country's conservation areas.

An underlying theme the INBio emphasized was Costa Rica's "Save-Know-Use" approach to biodiversity (WRI/IUCN/UNEP 1992). INBio has been a pioneer in working to make bioprospecting an asset for conservation. If practiced carefully, bioprospecting can be a nondamaging use of natural resources for two principal reasons. First, in areas set aside for the preservation of natural resources, successful bioprospecting can rely on an ecologically insignificant quantity of specimens, the same minimal level of collection that academic taxonomists collect as "voucher" specimens for museum work. Second, if the collection of, research on, and use of a collected specimen leads to a commercialized product, a portion of the proceeds can be earmarked for conservation.

5.2. Prospects

Although not directly transferable to a U.S. context, INBio's bioprospecting program for their conservation areas has many lessons for wildland management in the United States. The aim of the workshop was not to display Costa Rica as a model for the U.S.

National Park System, but rather to allow participants to examine how and where the Costa Rican experience is relevant to the United States. Officials of YNP, NPS, and the Department of the Interior (U.S. DOI) are currently examining several alternative arrangements to regulate and monitor microbiological research in the U.S. National Park System. Options include alterations of the permit system, Material Transfer Agreements (MTAs), "Special Use Agreements" at the Park Superintendent level, and Cooperative Research and Development Agreements (CRADAs) at the U.S. Department of the Interior (DOI, 1996) level.

6. EDUCATION

6.1. Current Status

To a degree, the process of educating the public about Yellowstone's microorganisms has had a strong start. The "Story of *Taq*" is a good story, filled with human interest, a fortuitous sequence of events, eccentric genius, and a scientific tool that is tangibly changing the social landscape (Rabinow, 1996). Not surprisingly, many conservationists have used the example of *Thermus aquaticus* as an example of how protecting public lands can lead to unexpected benefits. The possibility of prokaryotic life on Mars and the corresponding research in Yellowstone has also led to a sizable amount of publicity. In sum, during the past four years, many articles have addressed the issue of microbial diversity at YNP; many of these have focused on the wide range of public benefit that has accrued from protecting microbial resources (see Appendix II). From YCR staff to the NPS Director's Office, NPS officials are also working to educate the public about Yellowstone's thermophilic microbial diversity and its tangible benefits (Kennedy, 1995). Within the Park itself, visitors can learn about the microbial resources from a limited amount of Park handouts, purchasable pamphlets, and interpretive presentations on signposts. In total, however, it is unlikely that a typical visitor to Yellowstone will learn something, even something very basic, such as the fact that many of the thermal features' colors come from microorganisms, about Yellowstone's thermophiles.

6.2. Prospects

Public education about microbiology is difficult. As one microbiologist has noted, "everyone comes to Yellowstone to see bears and wolves, but few do; nobody comes to see microbes, but everyone does" (Niels Ramsing, personal communication). Compared to Yellowstone's grizzly bears, bald eagles, wolves, elk, and other "charismatic megafauna," microorganisms generally do not hold much attraction to the public eye. Educators have only begun to draw out the fascination of life in boiling water and highly acidic conditions, life where most of Yellowstone's visible creatures would quickly die (Lindstrom, in press). One potentially important educational tool is to relate microbial life to the aesthetic value of the colorful and highly visible features such as Grand Prismatic Hot Spring, Minerva Terraces, and Morning Glory Pool, among many others.

YCR along with the YNP interpretive staff is continuing to develop materials for public information. Ideally, most of the general information originating from the Park (including information disseminated in handouts and at visitor centers and interpretive displays) should contain substantively equivalent information on microbiology as on the

Park's geology and larger wildlife. In addition, accessible literature on microbial resources should be available as handouts or for purchase by Park visitors. In short, Yellowstone and the independent microbiologists who use the Park need to make Yellowstone's microbial resources understandable and recognizable to the general public.

7. SUMMARY

Each year, approximately 3 million people visit Yellowstone National Park to view its thermal features and wildlife. Few visitors are aware, however, that the thermal features themselves are home to an extremely diverse realm of wildlife: the biodiversity of microorganisms. Thermophiles are considered by some the most important species protected by the National Parks, and their potential to advance science and medicine has already been demonstrated. Academia and the biotechnology industry have sought out these microorganisms for what they can tell us about the evolution of life and for what they can contribute to biotechnological processes. *Thermus aquaticus* exemplifies success on both counts, allowing scientists to broaden our understanding of the microbial world exponentially while leading to the creation of an estimated $200 million per year industry. Undoubtedly, Yellowstone's thermophiles will continue to provide microbial ecologists and molecular biologists with a spectrum of genetic and phenotypic traits, ranging from heavy metal metabolism to the most primitive extant forms of life on Earth (Pace, 1991).

Exactly how the NPS should manage this nontraditional resource remains unresolved. But that is typical of Yellowstone, a Park that faces a host of controversial issues ranging from wolf reintroduction, fire policy, and bison management to overcrowding, snowmobile use, and development outside the Park's boundaries. In the case of managing its microbial resources, Yellowstone has a chance to avoid this type of conflict. Academic scientists, industrial biotechnologists, and government officials are all interested in increasing our collective knowledge of these enigmatic creatures. Some are especially interested in what these resources can tell us about the origins of life; others believe that these resources can benefit humankind in a many different applications. The NPS, on the other hand, is primarily interested in preserving these species and in protecting the rights of the owners of these resources, the people of the United States. Fortunately, cooperation between the academic, private, and government sectors could and should produce the advantageous symbiosis needed to study, develop, and preserve this cryptic resource.

ACKNOWLEDGMENTS. Thomas Brock, Norman Pace, Dave Ward, Robert Ramaley, and Anna-Louise Reysenbach have been staunch supporters of Yellowstone microbiology. In addition, Preston Scott and Leif Christoffersen have brought a balanced and rational perspective to the management of microbial resources. Without all of their help, microbial resources in Yellowstone would remain "the undiscovered resource."

APPENDIX 1: *THERMUS AQUATICUS* RECORD IN THE THERMOPHILIC MICROORGANISM SURVEY

Scientific name:	*Thermus aquaticus* (Taq)
Description:	Thermophilic, obligate aerobic eubacterium. Gram-negative, nonmotile nonsporing rods that produce a yellow cellular pigment, prob-

Description (*cont.*)	ably a carotenoid. Under certain cultural conditions, it forms long filaments and large spherical structures, probably related to spheroplasts. Generation time at optimum temperature is about 50 minutes. *T. aquaticus* was the first thermophile discovered and led to the science of life at high temperature; eventually, it yielded many other bacteria that grow at temperatures above 55°C.
Habitat type:	Neutral chloride Temperature range for growth: 40–79°C Optimum temperature: 70°C pH range: 7.0–8.0 Optimum pH: 7.5–7.8
YNP location:	Mushroom Pool, located in the White Creek area of the Lower Geyser Basin, off the Firehole Loop Road, 300 m east of White Dome Geyser, just inside the tree line. It is a circular pool approximately 20 m in diameter that has an L-shaped outflow channel and is covered with a dense green/brown bacterial mat.
First observed:	6/1/65
Other locations:	*T. aquaticus* is a common thermophilic microorganism with several strains found in hot springs around the world. It is also sometimes found in hot tap water in geographical locations distant from thermal springs.
Biotech use:	*Taq* DNA polymerase is the primary enzyme used in polymerase chain reaction (PCR), the enzymatic amplification of DNA for industrial use, which has revolutionized DNA science. The gene producing *Taq* polymerase has been cloned onto an *E. coli* host for mass production of this enzyme, a process patented by Cetus, Inc. using the original culture collected from Mushroom Pool.
Citation:	Brock, Thomas D. and Hudson Freeze. 1969. *Thermus aquaticus* genus nov., species nov.; a non-sporulating extreme thermophile. *J. Bacteriol.* **98**(1):289–297.
Photographs:	Photomicrographs in citation
Specimens:	Deposited in American Type Culture Collection as ATCC YT-1 25104

APPENDIX 2: BIBLIOGRAPHY ON THE MICROBIAL DIVERSITY AND COMMERCIAL USE OF MICROORGANISMS FROM YELLOWSTONE NATIONAL PARK (NONSCIENTIFIC LITERATURE)

Adams, Michael W. W., and Kelly, Robert M. 1995. Enzymes from microorganisms in extreme environments. *Chemical and Engineering News* **18** December, 32–42.

Bednarek, David I. 1993. Friction: Heat-loving bacterium roils two worlds. *Milwaukee Journal*, May 9.

Billings Gazette. 1993. Gazette opinion: Industries exploit first park. *Billings Gazette*, December 6.

Broad, William J. 1995. Clues to fiery origin of life sought in hothouse microbes. *New York Times*, May 9, B7/B10.

Brock, Thomas D. 1978. *Thermophilic microorganisms and life at high temperatures*. New York: Springer-Verlag.

Chester, Charles. 1996. Controversy over Yellowstone's biological resources. *Environment*, October, 11.

Clifford, Frank. 1994. Simpson case boosts microbe conservation. *Los Angeles Times*, August 31.

Guyer, Ruth L., and Koshland, D. E., Jr. 1989. The molecule of the year. *Science*, December 22, 1543.

Kennedy, Roger G. 1995. The fish that will not take our hooks. *Wilderness*, Spring, 28–29.

Lindstrom, Robert. 1996. Biodiversity, ecology, and evolution of hot water organisms in Yellowstone National Park: Symposium and issues overview. *Park Science*, Winter, 12–13, 19.

Madigan, Michael T., and Marrs, Barry L. 1997. Extremophiles. *Scientific American*, 82.

Milstein, Michael. 1993. Yellowstone's secrets: Prospecting in the park (special report). *Billings Gazette*, December 5.
Milstein, Michael. 1994. Yellowstone geysers spew fossil clues to primeval life. *Los Angeles Times*, December 14.
Milstein, Michael 1994. Molecular technique is hot stuff among microbe detectives. *San Diego Union*, June 1.
Milstein, Michael. 1994. Should park cash in on bacteria? *Billings Gazette*, January 10.
Milstein, Michael. 1994. There's gold in them thermophiles. *Outside*, August.
Milstein, Michael. 1994. Who has the right to "own"—and lease—a living thing? *San Diego Union*, June 1.
Milstein, Michael. 1994. Yellowstone managers eye profits from hot microbes. *Science*, 29 April, 655.
Milstein, Michael. 1995. NPS director: Nation must protect parks, lifeforms. *Billings Gazette*, January 2.
Milstein, Michael. 1995. Park wants money from its microbes. *Billings Gazette*, September 20.
Milstein, Michael. 1995. Primitive microbes found in hot springs. *Billings Gazette*, September 20.
Milstein, Michael. 1995. Research in park under scrutiny. *Billings Gazette*, January 2.
Milstein, Michael. 1995. Science sees potential of park "bugs." *Billings Gazette*, September 19.
Milstein, Michael. 1995. Yellowstone managers stake a claim on hot-spring microbes. *Science*, 13 October, 226.
Robbins, Jim. 1994. The microbe miners. *Audubon*, November–December, 90–95.
Wolf, Ron. 1994. Yellowstone discovery: Should U.S. get profits? *San Jose Mercury News*, July 25.

REFERENCES

Brock, T. D. 1978. *Thermophilic microorganisms and life at high temperatures*. New York: Springer-Verlag.
Brock, T. D., and Freeze, H. 1969. *Thermus aquaticus* gen. n. and sp. n., a nonsporulating extreme thermophile. *J. Bacteriol.* **98:**289–298.
Gelfand, D. H., Mullis, K. B., et al. 1988. Primer directed enzymatic amplification of DNA with a thermostable DNA polymerase. *Science* **229:**29.
Kennedy, R. G. 1995. The fish that will not take our hooks. *Wilderness*, Spring, 28–29.
Lindstrom, R. In press. Case study: *Thermus aquaticus*. In Vogel, J. H. (ed.), *From traditional knowledge to trade secrets: Prior informed consent and bioprospecting*. Washington, DC: InterAmerican Development Bank.
Madigan, M., Martinko, J., and Parker, J. 1997. *Brock biology of microorganisms*. Englewood Cliffs, NJ: Prentice-Hall.
Meagher, M., and Meyer, M. E. 1994. On the origin of brucellosis in bison of Yellowstone National Park. *Conserv. Biol.* **8**(3):645.
National Research Council. 1992. *Science and the national parks*. Washington, DC: National Academy Press.
Pace, N. R. 1991. Origin of life—facing up to the physical setting. *Cell* **65:**531.
Pace, N. R., Stahl, D. A., et al. 1986. The analysis of natural microbial populations by ribosomal RNA sequences. *Adv. Microbial Ecol.* **9:**1–55.
Rabinow, P. 1996. *Making PCR: A story of biotechnology*. Chicago: University of Chicago Press.
U.S. Department of the Interior. 1996. Technology transfer: Marketing our products and technologies: A training handbook for the U.S. Department of the Interior. Washington, DC.
U.S. National Park Service. 1993. Science and the National Parks II: Adapting to change. Washington, DC: U.S. Department of the Interior.
U.S. National Park Service. 1995. Natural resource inventory and monitoring in national parks. Washington, DC: U.S. Department of the Interior.
Ward, D. M., Ferris, M. J., et al. 1994. Species diversity in hot spring microbial mats as revealed by both molecular and enrichment culture approaches—relationship between biodiversity and community structure. In Stahl, L., and Canmette, P. (eds.), *Microbial mats: Structure, development and environmental significance*. NATO/AIS series publication.
World Resources Institute, World Conservation Union (IUCN), and United Nations Environment Program. 1992. *Global Biodiversity Strategy*. Washington, DC: World Resources Institute.
Yellowstone Center for Resources. 1996. Investigators' annual reports 1995. Yellowstone National Park: U.S. National Park Service. July. YCR Annual Report: YCR-IAR-96.
Yellowstone Center for Resources. 1996. Thermophilic microorganism survey. Yellowstone National Park: U.S. National Park Service. May. YCR Special Report: NRSR-96-1.

Index

MIX
Papier aus verantwortungsvollen Quellen
Paper from responsible sources
FSC® C105338

If you have any concerns about our products,
you can contact us on
ProductSafety@springernature.com

In case Publisher is established outside the EU,
the EU authorized representative is:
Springer Nature Customer Service Center GmbH
Europaplatz 3, 69115 Heidelberg, Germany

Printed by Libri Plureos GmbH
in Hamburg, Germany